GET
TECHNOLOGY

GERALD LYNCH

GET TECHNOLOGY

BE IN THE KNOW.
UPGRADE YOUR FUTURE.

WHITE LION
PUBLISHING

Brimming with creative inspiration, how-to projects and useful information to enrich your everyday life, Quarto Knows is a favourite destination for those pursuing their interests and passions. Visit our site and dig deeper with our books into your area of interest: Quarto Creates, Quarto Cooks, Quarto Homes, Quarto Lives, Quarto Drives, Quarto Explores, Quarto Gifts, or Quarto Kids.

First published in 2018 by White Lion Publishing
an imprint of The Quarto Group
The Old Brewery, 6 Blundell Street
London N7 9BH
United Kingdom

www.QuartoKnows.com

A catalogue record for this book is available from the British Library.

ISBN 978 1 78131 748 8
Ebook ISBN 978 1 78131 779 2

10 9 8 7 6 5 4 3 2
2022 2021 2020 2019 2018

Designed and illustrated by Stuart Tolley of Transmission Design

Printed in China

'Libraries gave us power.'

For Mum, Dad and the
Brotherhood of Lynch

CONTENTS

INTRODUCTION   08

HOW TO USE THIS BOOK   10

01 TECH BY YOUR SIDE

01 Virtual Reality   18
02 Augmented Reality   22
03 Artificial Intelligence   28
04 Smart Homes   32

→ Toolkit 01—04   36
+ Further Learning   38

02 EXPLORATION

05 Driverless Cars   44
06 Hyperloop   48
07 Exosuits   54
08 A New Space Race   58

→ Toolkit 05—08   62
+ Further Learning   64

03 SURVIVAL

09	Nanorobots	70
10	The Quantified Self	74
11	Nuclear Fusion	80
12	Asteroid Defences	84

| → | Toolkit 09—12 | 88 |
| + | Further Learning | 90 |

04 SECURITY

13	Cybersecurity	96
14	Biometrics	100
15	Blockchain	106
16	The Autonomous Army	110

| → | Toolkit 13—16 | 114 |
| + | Further Learning | 116 |

05 TRANSCENDENCE

17	Quantum Computing	122
18	Terraforming	126
19	Bionic Implants	132
20	Transhumanism	136

| → | Toolkit 17—20 | 140 |
| + | Further Learning | 142 |

| | Epilogue | 144 |

INTRODUCTION

Your world is changing.

Not since the advent of the agricultural revolution has mankind's actions given us the agency to fundamentally alter not only our daily lives, but our institutions, our planet and the very building blocks of life itself.

Where once a computer in every home seemed laughable, there's now a computer in every pocket – with some of the greatest minds around the globe looking to put a computer inside every brain too.

Hyper-accelerated growth in technological advancements is allowing access to information and welfare that would have seemed impossible little more than a decade ago, while also opening us up to new security and ecological challenges which, if left unchecked, could prove catastrophic.

With each passing day, our lives become ever more entwined with and dependent upon technology. And yet there's a feeling that we know less and less about the devices that we use, how they work or what their manufacturers are ultimately progressing towards.

This book looks to change that. It aims to act as an introduction to some of the most exciting and important technological advancements and concepts currently being developed or theorized upon at this very moment in history. Some chapters will tackle the cutting edge of what is available today, describing gadgets that you could order right now with the tap of a smartphone app, while others discuss ideas that may not come to pass in your children's children's children's lifetimes.

All, however, will show you not only what the scientists and technicians of the world are looking to build, but what these devices, algorithms and ways of thinking will let us become.

With each passing day, our lives become ever more entwined with and dependent upon technology.

HOW TO USE THIS BOOK

This book is organised into five parts and 20 key lessons covering the most current and topical debates of technology today.

Each lesson introduces you to an important concept,

and explains how you can apply what you've learned to everyday life.

As you go through the book, TOOLKITS help you keep track of what you've learned so far.

Specially curated FURTHER LEARNING notes give you a nudge in the right direction for those things that most captured your imagination.

BUILD + BECOME

At BUILD+BECOME we believe in building knowledge that helps you navigate your world. So, dip in, take it step-by-step, or digest it all in one go — however you choose to read this book, enjoy and get thinking.

KNOW TEC
TODAY TO
YOURSELF
FOR TOMO

HNOLOGY
EQUIP

RROW.

TECH BY
YOUR SIDE

LESSONS

01 VIRTUAL REALITY
Will virtual reality open up a universe of possibilities to everyone?

02 AUGMENTED REALITY
Can augmented reality change the way we see — and interact with — the world?

03 ARTIFICIAL INTELLIGENCE
Separating AI fact from fiction — should we fear an autonomous robot takeover?

04 SMART HOMES
How will smart tech and even smarter engineering make the homes of the future work for the planet — and for you?

The science fiction of yesterday is poised to become the routine of tomorrow.

Smart homes, artificial intelligences (AIs), robot doctors – it's hard to keep up with the fantastic advances seen in our everyday reality, before even considering throwing 'augmented' and 'virtual' realities into the mix.

But with the use of technology now so ubiquitous in all areas of society, change comes quickly, and its impact can have great repercussions for all walks of life. From the student given an entire universe to explore through a VR headset to the factory worker having to rethink his place in the world alongside a robot colleague, technology has just as much potential to expand our understanding of the world as it does to ruin our usefulness within it.

The science fiction of yesterday is poised to become the routine of tomorrow. Are we ready for it? How will we entertain and educate ourselves with devices that offer infinite possibilities? How will we harness the increasing flow of data we're constantly bombarded with? And is the AI apocalypse of so many Hollywood horror stories really on the horizon?

This chapter will discuss the technologies set to sit alongside us as we learn, live, work and play in the coming years. You'll use some in the office, wear others in the street and be greeted by others as you walk through your front door. Many exist already in their infancy, and understanding where they are now (and where they're heading in the future) will be key to keeping pace as they alter the way we live and run our lives.

VIRTUAL REALITY

You will never visit Jupiter. You'll never climb Mount Everest solo, you'll never experience life in ancient Rome and you'll never walk with dinosaurs. At least, not in this reality.

We're living through the early stages of a revolution, one in which technology is on the precipice of democratizing access to experiences considered to be out of reach for the majority of people. By equipping users with vision-enveloping headsets, powered by increasingly capable computer processors, virtual reality (VR) software manufacturers will be able to take us on incredible journeys, anywhere on or off the globe (or to places wholly imagined) from the comforts of our homes.

Current virtual reality headsets broadly take two forms. The first are mobile VR viewers, which are battery powered and either make use of an inserted smartphone or embedded computing, sensor and display components. The second, currently more advanced, form of virtual reality makes use of a tethered headset hooked up to a powerful PC or games console, and can include remote sensors to allow a user to be tracked in 3D space. While mobile headsets offer greater flexibility and freedom of movement, they can't yet compare to tethered headsets like the HTC Vive or Facebook-owned Oculus Rift when it comes to conjuring the so-called

feeling of 'presence' – that sense that you're truly inhabiting a realm beyond the one you are physically occupying.

Both are based on similar underlying technologies, and their core components serve essentially the same purpose. To deliver a convincing virtual reality experience, VR headsets must match visuals on screens in front of your eyes with the pitch, yaw and roll of your head, as well as shifting movements in directions up, down, left, right, backwards and forwards. In some cases, they may have to do this while tracking your limbs too, in a way that doesn't make your appendages feel disembodied. To do this, they use:

01. Sensors Magnetometers, accelerometers and gyroscopes all work in tandem. Magnetometers measure magnetic fields, and can orientate the headset relative to magnetic north, helping the processor establish a fixed point of reference with which to position the user and experience. Accelerometers can determine which way up a device is, and when set up in a series can measure acceleration along an axis, an element used in tracking a VR participant's movement. Gyroscopes are also used to measure orientation, and can be used to match a VR headset-wearer's head movements to the unfolding action.

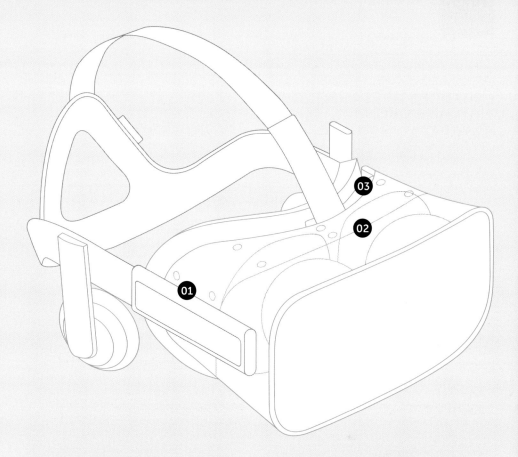

02. High-resolution displays The more pixel-dense these are, the better. In order to be immersive, virtual reality experiences require sharp screen visuals and an extremely fast refresh rate (how quickly the display updates its image to reproduce smooth motion). Anything less than 90Hz (90 refreshes per second) is linked to detrimental feelings of nausea and motion sickness, where a VR participant's brain can't marry the body's movements to the images it is seeing. Virtual reality headsets use binocular-like goggles in front of the screens to deliver stereoscopic 3D visuals.

03. Processing units Whether embedded within the headsets themselves, or located remotely in an inserted smartphone or tethered PC, central processing units (CPUs) and graphical processing units (GPUs) are required to bring to life the software visuals that transport a virtual reality explorer to another realm. These need to be relatively powerful – rendering complex 3D visuals at high definition and at a fast refresh rate is a demanding task.

ENHANCED LIVES THROUGH VIRTUAL WORLDS

With the first wave of virtual reality headsets now in homes, we've already had a glimpse of the potential that VR experiences can offer. Want to take a dive in a shark cage? Sony's PlayStation VR headset will let you do just that. Want to tour the world from the comfort of your living room? Pop on Google's Cardboard VR viewer and marvel at 360-degree views of the world's greatest landmarks. Want to make a digital sculpture while floating in space? Strap yourself into a HTC Vive and fire up the Tilt Brush application.

What about intercontinental meetings conducted via an avatar, a virtual you with all the minutiae of gestures and body language we experience now through face-to-face interactions? You'll be able to generate the setting for the meeting too – why talk through the month's sales figures in the boardroom when everyone would find it much more enjoyable to do so on a digital beach? Such freedom will have the potential to build all-new global communities, with each inhabitant offered the illusion of genuine proximity to their neighbours.

This will raise important questions over identity and the self within virtual worlds. When you can program your virtual-reality avatar to look however you desire it to – potentially massively removed from your real-world body – what will that mean for interpersonal relations? When all elements of the imagination can be explored at will, would it even matter? When an entire alternate reality is built from scratch, how many of the real-world societal norms will we want to keep in it?

The potential for virtual reality is vast, and seemingly limitless. There may come a day where the line between your physical reality and the virtual one is indistinguishable. The possibilities are as overwhelming as they are breathtaking.

Educative VR offers the potential to scale experiences right from pre-school to advanced studies. For instance, it would be incredibly expensive to fly a child from the USA to the Louvre in Paris to see the Mona Lisa, but for a few hundred dollars an entire class could walk the gallery's halls in VR. Or perhaps VR could allow a veterinary major to marvel at the intricate ligament structures inside a horse's leg without having to dissect a cadaver.

All these experiences will become even more engrossing once we've perfected haptic interaction systems – interfaces that will allow us to feel the digital world around us. In the future, we may wear full body suits that let us feel the most tender of caresses or minute changes in temperature in a software environment, and make use of gloves that would allow a virtuoso pianist to perform an intricate piece on a piano made of bits and bytes to an audience of millions, all with front-row seats.

AUGMENTED REALITY

You may not realize it, but you've probably already used an augmented reality (AR) application. Did you jump on board the Pokemon Go mobile app game craze? If you spotted a Pikachu walking down your street through your phone, that was an augmented reality moment. Likewise, if you've put an animated 'Sticker' on a video in Snapchat, such as popping a wagging dog's tongue on a friend's well-posed pout, that too was using AR systems.

But this is just the start. While the screen may never go away entirely, augmented reality advances will lead to devices that push the screen to the periphery of your vision, turning displays into digital windows and making virtual elements appear to exist in the real world around us.

Google tried it with Glass, and iPhone-maker Apple is investing heavily in this technology, but it's fitting that it is Microsoft – gatekeeper of the Windows operating system platform, that most ubiquitous of workplace computer software – that's leading the charge in augmented reality hardware.

In fact, Microsoft is so confident of the growing capabilities of its HoloLens headset that it refers to it not as 'augmented reality' hardware, but as 'mixed reality'. It makes a great case study for what to expect from the AR gadgetry of the future.

Where HoloLens and its imitators really separate themselves from virtual reality headsets is through their image-recognition capabilities. In conjunction with motion-tracking sensors and an internet connection, an ambient light sensor and four environment sensing cameras work to track and map your surroundings. With the depth calculated (and the cameras keeping an eye on your limbs too), you'll be able to walk up to AR 'holograms' and 'touch' them.

At its most basic, you'll be able to take your desktop computing applications and throw them onto the walls of your home on a grand scale. But AR gets most exciting in the outside world, when all the sensors and cameras work together to feed you information on your surroundings in real time, without being prompted.

Like the VR wearables of the previous chapter, HoloLens can be worn on the head, and features sensors and an on-board processing unit. It too has a screen that sits in front of your eyes, displaying digital graphics. But whereas VR headsets envelop your view, transporting you to other worlds, HoloLens has a transparent screen, overlaying the digital software elements onto your actual surroundings in front of your eyes. This gives the graphics a sense of weight, depth and presence in the real world, ready to be manipulated and interacted with.

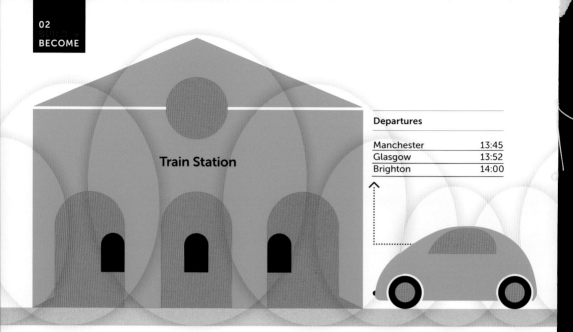

Departures

Manchester	13:45
Glasgow	13:52
Brighton	14:00

Train Station

DEMOCRATIZING EXPERTISE

The modern workplace is dominated by the keyboard, the computer monitor and the cubicle. The personal computer, laptop and fibre internet liberated us from the office, but they didn't release us entirely.

Augmented reality offers fantastic possibilities to shake up expectations around how we work, play and live. With an AR system strapped to our heads and our hands free to interact within both the physical and graphically rendered realms at the same time, our relationship with both the world around us and the digital one will fundamentally change.

Suddenly, an architect can visit an empty plot of land and, with the wave of a hand, pull up digital skyscraper schematics and see where they will exist in the real world, as if towering into the clouds. A rocket scientist can take her latest propulsion system from a desk-sized model and out onto the launchpad without burning a drop of fuel, and an international student can have his new overseas textbook magically translated into a language he understands – at least through an augmented reality lens.

AR devices will also allow humans, for the first time since the age of the hunter-gatherer, to become generalists again. Whether you're a car mechanic or a sushi chef, you have a very specific skill set unique to completing tasks within your chosen profession. But a person wearing an augmented reality visor, with access to the right applications, could give even the professionals a run for their money. Sure, popping on a computer won't be a replacement for years of learned experience, or turn you into an instant grease monkey. But if you can look through an AR lens at a dismantled car engine and (provided all the tools are at hand) be given visual, step-by-step instructions overlaid onto the real world as to which piece goes where, you'd suddenly have far more confidence in copying the steps to complete a task unaided.

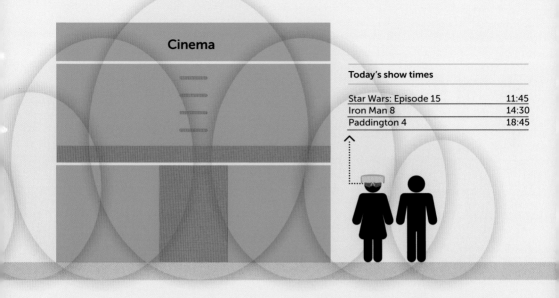

Cinema

Today's show times

Star Wars: Episode 15	11:45
Iron Man 8	14:30
Paddington 4	18:45

Everyday tasks get a mixed reality boost too. Imagine a journey home from work while wearing a pair of AR spectacles. You pop into a supermarket, unsure of what to have for dinner — you pick up a packet of mushrooms, which the glasses recognize and suggest a few recipes around. If the supermarket has been mapped, the specs may even direct you around the aisles to help you find the items you need. You then race to the nearest train station, but a combination of Global Positioning System (GPS) sensors and image recognition means that, when looking at the station entrance, you're presented with information revealing that there are severe delays on your route home. A voice command requesting a way to kill some time results in a trailer for a movie you've shown interest in, placed in the corner of your peripheral vision — and it happens to be showing in 10 minutes' time at a cinema just a street away. With the directions then quite literally traced at your feet, your journey home has taken an unexpected turn — even if that mushroom risotto has to wait.

AUGMENTED WILL FUNDA CHANGE THE INTERACT W THE WORLD.

REALITY
MENTALLY
WAY WE
ITH

ARTIFICIAL INTELLIGENCE

From *2001: A Space Odyssey*'s chilling AI watcher HAL to the *Terminator*'s doomsday-triggering Skynet, artificial intelligence has been presented as the technological bogeyman for decades. You'd be forgiven for misunderstanding the term — no two words have been more often misused, with 'AI' slapped on the box of many a product with barely a flicker of intelligence of any kind present. For years, 'artificial hype' may have been a more apt description.

So what is the sentient or 'general' artificial intelligence that science fiction has made us so greatly fearful of? What is so-called 'narrow' AI? Take a smartphone's voice-activated assistant. Though your iPhone's Siri helper may appear intelligent, it's not capable of actual thought — at least beyond the specific functions machine learning and deep learning have made it possible to program into it. It's an example of 'narrow' AI, or more specifically, a hybrid AI that takes a group of 'narrow' AI abilities (speech recognition, primarily) and uses them to tap into the huge reams of constantly updating data in the cloud to give the illusion of superhuman intelligence.

Just because Siri can instantly access all the knowledge of the internet, that doesn't mean it's capable of creative thought or reason — it's simply completing a task it has been programmed to do very ably. Only when Siri could become capable of carrying out any action that a human could perform could it be considered a true AI. And no AI has ever been able to do that — at least yet.

So how far-reaching will AI be? There are two schools of thought here. The first comprises those that fear the rise of a superintelligence — a group that counts Bill Gates and Stephen Hawking among its ranks. They foresee a day where the advent of a 'general' AI leads to an 'intelligence explosion' — a moment that would see AIs capable of teaching themselves new things at a runaway rate and eventually surpassing human capabilities to such a degree that they ultimately enslave us. This grim vision of the future, known as the 'technological singularity', is offset by the fact that it is a philosophical fear as much as a practical one — there exists no empirical evidence towards an AI reaching this point. But likewise there is nothing to prove irrefutably that it couldn't happen.

The second group sees the advance of AI in far more pragmatic terms. The intelligence explosion won't occur, but the 'narrow' AI we use today will continue to dominate. That's not to say 'narrow' artificial intelligences won't become amazingly useful. Their capabilities will grow exponentially, and though they won't become capable of true, creative thought, they'll become incredibly skilled and efficient in hyper-specific roles, where context is contained and nuanced interpretation unnecessary. It's here where the converging worlds of AI and robotics will usher in the most change for humans.

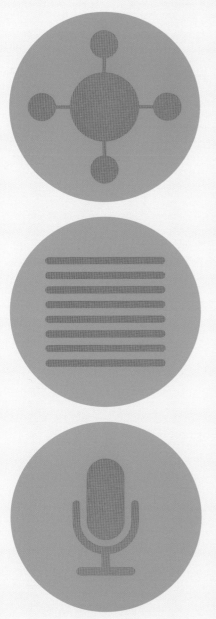

AI builds on the advances of a combination of computing methods:

01. Machine learning Algorithms sift through huge quantities of data to learn about its contents in the context of a specific need and make a decision to complete that task. The idea is to allow the algorithms to do the heavy lifting, autonomously trying different approaches to a task before hitting a brick wall, after which a coder usually still has to step in to create classifiers to refine the process.

02. Deep learning This process uses neural networks, which are designed to more closely resemble the way humans think, attempting to mimic the neurons of our brains. Essentially, it allows a task to be considered from multiple angles at once – building layers of related information about a problem in order to find a solution. Each layer then weighs up the possibility of having identified its specific element correctly, before feeding all this information to a final layer that ultimately has to choose the answer. It takes a huge amount of processing power and large data sets, and at this stage is a slow process.

03. Big Data The reams of information we generate, store, transmit and share through our devices and their sensors every day, and the way they can be used collectively to find correlations and build patterns of behaviour, are known as Big Data. Software, combined with computer GPUs that are incredibly adept at carrying out the parallel processing cycles required to make sense of that data at high speed, make Big Data very useful to AIs.

IS IT WISE TO FEAR AI?

We already see robots capable of carrying out simple repetitive tasks like those required on a factory production line. But the cultivation of ever-improving AI will accelerate AI adoption and lead to robots capable of increasingly complex actions.

This is where some apprehension around AI is valid. Artificially intelligent robots, even those powered by 'narrow' AI, will become so efficient at some tasks as to make them increasingly more cost effective to employ than human workers. AI-powered machines are most likely, at least initially, to begin to take over low-skilled jobs such as truck driving, production line work and customer-facing retail roles.

As a result, there are many questions the future will need to ask about how best to integrate artificial intelligences into

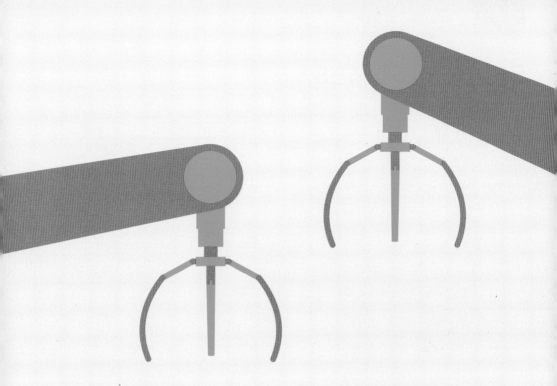

society. Will we have to find a way to tax the robots that have taken these roles – or the multinational corporations that install them? Is a universal basic income a necessity our governments need to factor into future budgets, and will there need to be a more robust way of tackling the 'skills gap' that leaves some members of society unable to support themselves?

The AI revolution isn't the first technological upheaval that the workforce has weathered and eventually co-existed with. Contrary to fears at the time of its arrival, the rise of the computer didn't lead to the loss of all our jobs, and instead led to whole new supplementary industries being formed. Could a job seeker in the 1940s have predicted the booming video game industry's need for digital artists? Or the rise

of the smartphone app programmer? Or the pop-chart-baiting YouTube star? Without computers, none of these professions would exist. Likewise, AI-driven robotics will almost certainly lead to the appearance of all-new, as-yet unimagined jobs and industries.

It is also important to note the potential that ever-improving AI-driven robots will have to enrich our lives. Think of the tireless robot manning a care home through the night, or the AI-charged robot teacher with eternal patience and an answer to every question. Picture a disaster scene like that of the Fukushima nuclear meltdown – its clean-up exposed hundreds of people to life-threatening radiation levels. What if a smart, AI-powered robot could have taken on their roles? Remember there are times when humans don't want to get their hands dirty.

SMART HOMES

When you think of your home, you probably think of its bricks and mortar, your favourite squidgy armchair and maybe your prized flowers in the garden. But it's easy to forget that beneath the carpet and behind the wallpaper lives a complex set of interlocking systems. From the electrical wiring to the plumbing, these circuit breakers and pipes ensure you live in relative comfort.

The way we approach home-building is constantly evolving, based on the needs of the society of the time. Two trends are set to change the way we live once again – the proliferation of convenience-driven smart devices, and the green-championing build that produces more energy than it consumes.

Smart devices have been quietly invading our homes for the last decade. As Wi-Fi became ubiquitous, broadband faster, cloud computing services common and sensors and chipsets cheaper, manufacturers set about producing networked variants of many common household items. You can now walk into a shop and buy a Wi-Fi-connected colour-changing light bulb that's controlled by your smartphone, a video doorbell that sends images to your phone if you're away from your property, and even air-quality monitors that alert you to when you should be taking a hay fever pill.

This so-called 'Internet of Things' is being pursued by all the major Silicon Valley companies, as they each look to create the one system to rule them all. In exchange for data on the minutiae of your home life, Google's Assistant, Amazon's Alexa and Apple's Siri-powered HomeKit will act as the

For the smart home to achieve maximum efficiency, it will need to have been built (or retrofitted) to produce more energy than it consumes. This means building your home in such a way that (thanks to the use of solar arrays and systems alongside passive heating principles) the appliances you use not only have their net annual carbon emissions from fuel burning reduced, but that the building exports leftover stored green energy back to the grid for personal profit. While you may still need to import energy into a carbon-positive home during the winter months, careful build planning – such as the use of insulated render, solar-powered air heating systems, energy efficient appliances and an educated approach to personal energy use – will mean that the summer months will be an energy gold rush for some homeowners.

It's far easier to put these principles into practice in purpose-built new builds than it is to rework older properties. But even adhering to just some of the ideas of the smart home will have both environmental and financial benefits in the long run. Homes of the future will make those of the last century look like ecological vandalism.

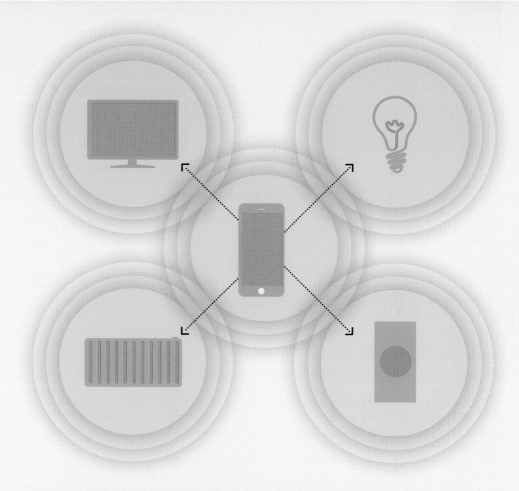

The next step is to tie all these devices together so that they talk to each other, and adapt their services based on your needs and the readings of each separate device.

bridge between all your connected devices, letting you control each or all at once with just a voice-activated command.

Once networked, these devices will work in tandem to bring greater convenience to your everyday actions, using 'narrow' AI techniques to learn your habits and tastes, and bring greater efficiency to the interdependent systems of your home.

You'll return in midwinter from a day at the office to find your networked thermostat has warmed the house to just the temperature you like (and not a degree more), the lights are on in your entry hallway, your favourite chillout album is playing from your networked stereo, and the fridge has ordered the milk you were running low on.

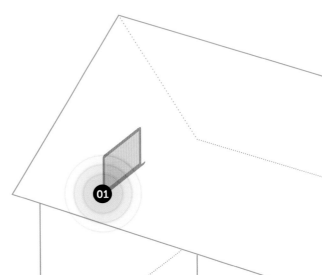

01

A GREENER, RICHER FUTURE

Home is where the heart is, but in the future it may also be where you top up your salary. With space in our cities at a premium, climate change creating significant problems for future generations and the price of energy set to fluctuate as fossil fuels become more costly to extract, we're going to have to become far more resourceful when it comes to designing and building our future homes.

Smart tech and clever engineering will come together to ensure we squeeze every possible kilowatt-hour from our energy supplies. With sensors communicating with generators, thermostats learning your commuting patterns and motion sensors making sure a room light only stays on when you're by it, we'll minimize our draw from the power grid. And, by kitting out our environments with green energy harvesters, we'll be selling energy back into the grid, perpetuating a greener way of living – and putting a tidy chunk of change into our pockets in the process. We'll be living the green dream. The same principles will be applied to the way we purchase and consume our foods. UK households threw away £13 billion worth of edible food in 2015 – the equivalent of £470 per household that year. This waste will generate substantial avoidable greenhouse gases as it degrades. But with a smart fridge monitoring what's left inside it, suggesting recipes for leftovers and cautioning against unneeded items on your smartphone shopping list, we'll be able to keep grocery costs down and make sure good food ends up where it is supposed to – in our bellies.

It's the way many of these energy-saving processes will be carried out without our direct daily input that makes the promise of the smart home so appealing. Most people like to think they do their part to help the environment, whether by recycling or taking a shorter shower in the morning. But the moment a green activity starts becoming inconvenient, we tend to lapse back into bad habits. The smart home ideal looks to strip the hassle away – you do your part just by buying into the concept and living with its systems, while the tech does the hard work of making good on your best intentions.

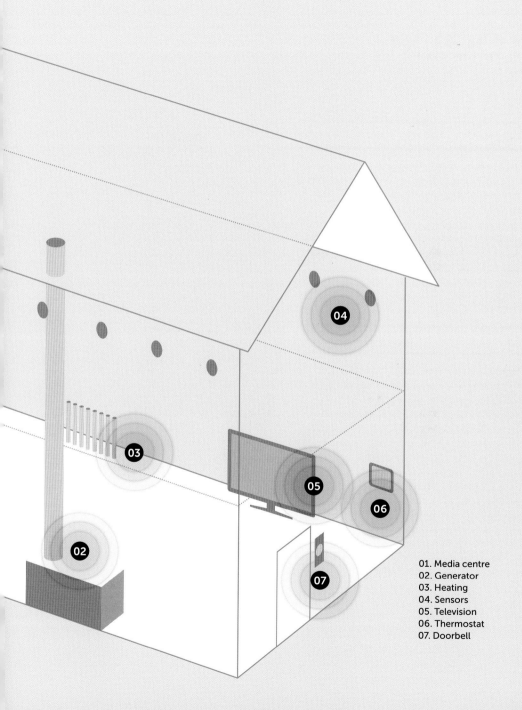

01. Media centre
02. Generator
03. Heating
04. Sensors
05. Television
06. Thermostat
07. Doorbell

TOOLKIT

01

Virtual reality will allow people from all walks of life to experience sights and sounds from around the globe – or a world that is wholly imagined. Having a physical presence in digital worlds, we'll build new communities in virtual spaces, and interact with each other and our rendered surroundings using haptic feedback wearable technologies.

02

Augmented reality will take the digital information we're so used to interacting with on screens and project it onto the world around us. AR headsets and glasses will intelligently present us with contextually relevant facts, figures and direction, giving us the confidence to try new activities and experiences without direct tuition. It will also free us from the office cubicle and computer monitor, turning our environments into mixed-reality workspaces.

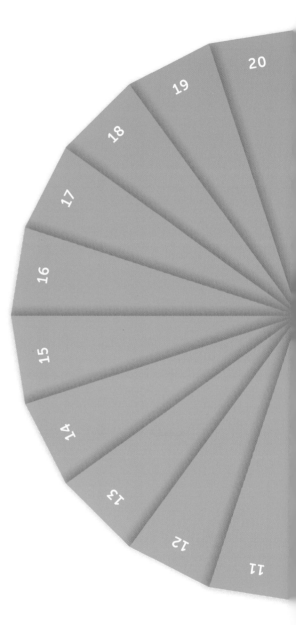

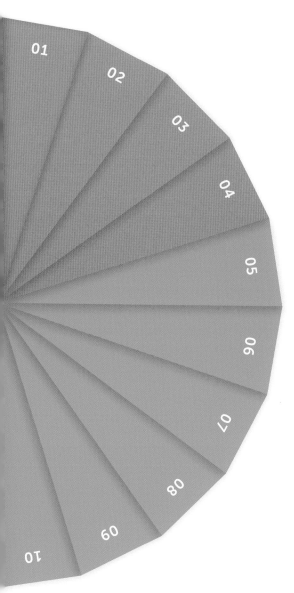

03

Artificial intelligences will continue to advance to a point where they will increasingly become more capable than humans at many tasks. AI will aid us in many areas of everyday life, and will lead to increased robotic automation across industries. But it will be a long time before they overthrow us entirely – if ever.

04

Smart devices will make our homes run like clockwork, managing everything from our lighting to our larders to make our lives easier and our homes greener. Coupled with building techniques that incorporate renewable energy-harvesting systems, we'll have the chance to live in green buildings that sell energy back to the grid, turning a profit for their owners.

FURTHER
LEARNING

READ

Snow Crash
Neal Stephenson (Penguin, 1992)

Neuromancer
William Gibson (Gollancz, 1984)

**The Master Algorithm: How the Quest
for the Ultimate Learning Machine Will
Remake Our World**
Pedro Domingos (Penguin, 2015)

**Practical Augmented Reality: A Guide to
the Technologies, Applications and Human
Factors for AR and VR**
Steve Aukstakalnis
(Addison-Wesley Professional, 2016)

**The New Net Zero: Leading-Edge Design
and Construction of Homes and Buildings
for a Renewable Energy Future**
William Maclay
(Chelsea Green Publishing, 2014)

WATCH

Meet Henry
Though there are many examples of VR
filmmaking you can experience, this Emmy
award-winning animation has charm to
spare. It puts you in the centre of the action,
though it does currently require an Oculus
Rift headset to view.

STUDY

**Electrical Engineering and Computer
Science**, Massachusetts Institute of
Technology, USA

Architecture, Kingston University London, UK

VISIT

VR Fest, Las Vegas, Nevada
This annual event takes place each January,
and shows the cutting-edge work being
done in the virtual reality space.

If augmented reality is of more interest to
you, simply visit your local Apple store — it
will have plenty of staff waiting to show you
the latest AR features of iPhones and iPads.

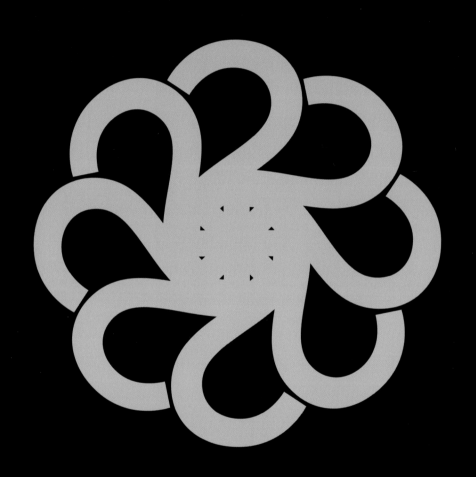

EXPLORATION

LESSONS

05 DRIVERLESS CARS
What's inside a driverless car, and how will these vehicles change our lives and cities?

06 HYPERLOOP
Cars, boats, planes and trains — will Hyperloop become the 'fifth mode of transport'?

07 EXOSUITS
How will technology augment human capabilities beyond the natural limits of our bodies?

08 A NEW SPACE RACE
Setting our sights on the stars, how are we taking our first steps towards a mission to Mars?

Technology is once again set to push our ability to explore unknown realms in exciting new directions.

From the voyages of Christopher Columbus to today's humble package holiday, humans have always craved adventure. But adventure is rarely found on our doorsteps. Travel – by cart, car or cruise liner – has always played a part in our grand expeditions.

Whether it was the wheel or the rocket, there has always been a technological or engineering breakthrough that has enabled us to take first steps into uncharted territories. And technology is once again set to push our ability to explore unknown realms in exciting new directions, rebuilding the ways we travel on our own planet, and setting new distant targets in our solar system.

This chapter explores the modes and methods being employed to allow us to take the next grand exploratory steps for our civilization – from the transformative commercial forms that will ease our daily commutes and empower us to hunt fresh opportunities beyond our atmosphere, to the personal technologies that have the potential to allow those that have never been able to bear their own weight to climb mountains.

We explore the boundaries of our known universe for many reasons – be that a hunt for resources, knowledge or meaning in a chaotic world. But the ultimate goal remains the same – to improve the lives of those living now, and provide a path for future generations to follow. Technology is leading the way. We should do our best to keep up.

DRIVERLESS CARS

Picture a motorway with three vehicles on it, each travelling at 100mph (160km/h). One car has a drunk man seated in the front seat, dozing, the second vehicle (a truck) is filled only with livestock, while a car hurtles along with no living passengers at all.

It sounds like the scene of a tragedy waiting to happen, but it is in fact a reality the automobile industry is hurtling towards.

Automated, driverless vehicles are leaving the drawing board and beginning to hit the tarmac across the globe, bringing about a revolution in road safety and transport efficiency.

Everyone from traditional vehicle manufacturers like Ford, BMW and Nissan to relative transportation upstarts like Google and Tesla are taking the concept for a spin. It's a moment of synergy where two once-separate worlds – those of the motor industry and Silicon Valley – will collide in breathtaking fashion, making a sci-fi dream come true.

The maturity of wireless technologies, intelligent camera arrays and versatile sensors will allow us to take our hands off the wheel and enjoy the benefits of a computer-controlled chauffeur.

As you can probably tell, few of these systems are unique or brand new. What is innovative, however, is the way each will come together, working in unison to deliver a whole new transportation method. In many respects, driverless cars could happily take to our roads and motorways tomorrow, were it not for us. The biggest challenge they face is human interaction – humans react erratically in manually operated cars.

Likewise, current road infrastructure is built with human senses in mind – from stripy school crossings to colour-coded motorway exits. A road system filled only with driverless cars could strip away all this furniture, relying solely and comfortably on inter-car communications, were it not for the millions of human car-drivers that depend on them.

While the external body design of driverless cars could vary from single-person bubble pods to lengthy commercial convoys, the underlying systems will remain more or less the same. A driverless car will need to make use of:

01. Wireless technologies In order to determine a destination, driverless cars will use combinations of GPS tracking and next-generation '5G' mobile internet networks. 5G, eventually offering speeds 12 times faster than current 4G standards, will also have the important job of letting driverless cars communicate with each other in order to anticipate each other's movements.

02. Video cameras and image-recognition systems Able to discern colours and shapes, video cameras will be able to spot and read traffic lights and road signs, as well as detecting pedestrians and other potential obstacles.

03. Radar sensors These will be placed around the car to track the position of nearby vehicles.

04. Ultrasonic sensors Used for moments when minute precision is required, ultrasonic sensors will come into play when squeezing into tight parking spots or pulling close to a curb.

05. Lidar sensors These sensors bounce light off lane markings and road edges to let the car understand its position in a lane.

06. Processing computer Taking in all the information from the previous sources at an imperceptible speed, the processing computer will control the steering, acceleration and braking appropriately.

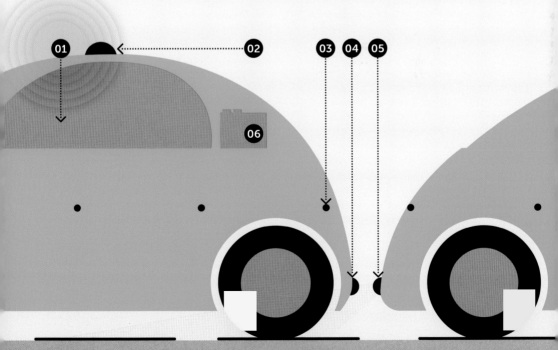

Efficient

THE AUTOMATED HIGHWAYS OF TOMORROW

Driverless tech has potential far greater than merely taking the strain out of a lazy trip to the shops.

The first and most obvious boon of the driverless revolution will come with road safety. Even the most experienced human driver has an off day – perhaps it was a boozy late night before a lengthy drive, or a wasp flying through the driver's side window on a hot day. These moments of tiredness or distraction can be fatal. But a computer never gets tired, and couldn't care less about a potential wasp sting. Networked together, driverless cars will be able to move in harmonious union, keeping each other informed of safe braking distances and potential obstacles, and never slipping into the road-rage-induced risk-taking that gives the White Van Man such a bad reputation.

Networked communication between cars could have a positive impact on the environment too. When all cars know the positions of each other, finding the most efficient route to a given destination will be simple – congestion will be reduced and, on those occasions where time is not of the essence, speed can be regulated to ensure the maximum gain is drawn from any chosen power source.

It's no coincidence that driverless cars are predominantly being designed with electric motors in mind, as vehicle emissions increasingly become a point of legislation that car manufacturers have to contend with. For human drivers, the relative scarcity of charging points can be a turn-off when given the choice of purchasing an electric vehicle. But an idle networked driverless car, always

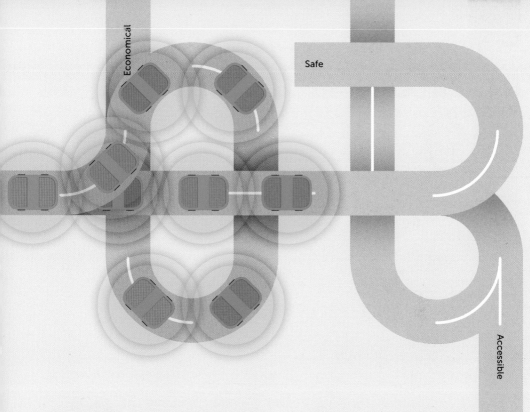

Economical

Safe

Accessible

aware of its surroundings, could be left to pop off to its local charging station without its owner being present, ensuring it's ready to go when needed again.

The economic benefits can't be undersold, either. Shipping companies will be able to take advantage of automated convoys, delivering large quantities of goods at speed without the need for rest breaks or traffic-induced delays. Public transport systems could run with never-before-seen efficiency, while the strain would be taken off arduously long commutes.

And best of all? The driving test, which inspires so many sweaty brows and sleepless nights, would also become a thing of the past. The freedom of movement that Henry Ford's Model T automobile first promised more than a hundred years ago would finally be fully realized.

What's to say a driverless car even needs a fixed owner? As society becomes more comfortable with the so-called 'sharing economy' (think services like Airbnb, which lets you rent a fellow user's home when vacant, or the many car- and truck-hire apps now available), driverless cars will encourage green-friendly carpooling journeys.

It doesn't take a great leap in imagination to foresee a service that can alert you to a driverless car ready for hire in your area, filled with passengers heading to a common destination.

HYPERLOOP

Cushion of Air

Boats, trains, cars and planes – for a long time these four forms of transportation have allowed humans to explore and conquer the world.

But as the world becomes ever more populous, the speed of industry intensifies and our existing infrastructure appears increasingly incapable of keeping up with the frenetic pace of modern life, the need for an all-new 'fifth mode' of transportation is becoming apparent.

Billionaire entrepreneur Elon Musk, founder of payment service PayPal, electric supercar company Tesla and commercial space-travel firm SpaceX, believes he has the answer: Hyperloop.

Musk intends to place passenger pods inside closed tubes which have had their air pressure reduced to one-sixth that of the thin air pressure on Mars. This would reduce drag on the pods, allowing them to travel at near supersonic speeds. The tubes would be elevated off the ground on concrete pylons as tall as 30m, or buried in deep tunnels, allowing for relatively flexible placement along built-up routes, and reinforced to withstand natural disasters like earthquakes.

The passenger and cargo pods would sit on magnetic metal skis, levitating on a cushion of air pumped through holes in the skis, generated by the pod's momentum through the tube. The pods would glide much like the pucks on an air hockey table by using electromagnetic propulsion, with an electric compressor at the nose of each pod pushing even more air backwards. Magnets on the skis would take an electromagnetic pulse jumpstart for a sensation similar to a plane on lift-off, after which you'd be travelling towards a theoretical top speed of 760mph (1,225k/ph) – just 8mph (13k/ph) shy of supersonic speed. That would take you from London to Edinburgh in just 45 minutes, or from San Francisco to Los Angeles in just 30! The first commercial Hyperloop route, connecting Dubai to Abu Dhabi, would see journey times of just 12 minutes.

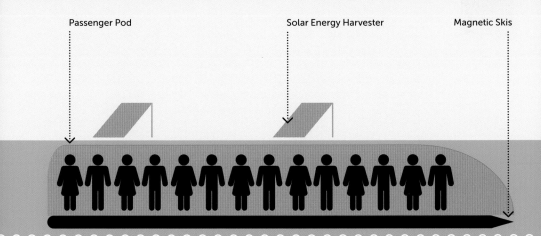

Passenger Pod   Solar Energy Harvester   Magnetic Skis

Solar energy harvesters would also be in play, covering the length of the tubes, while wind and geothermal energy generated inside the pipes could also be collected. Even the pod braking systems hold potential energy-storage capabilities. In fact, the Hyperloop team believes enough energy could be stored not only to make the system self-sustaining, but actually to generate power for the grid too.

It would be a little bit claustrophobic for passengers – but not dramatically less comfortable than, say, cities' subway systems. Pods would have a maximum height of 1.10m, and a width of 1.35m, with 28 passengers per pod sat in a row of pairs. At rush hour, Musk envisions a pod leaving every 30 seconds and, with tickets estimated to cost no more than $20 (£15), it would be a genuinely affordable alternative commuting route. That is, providing you don't need a toilet break while strapped in for the duration of the ride – no one's come up with a solution to that problem just yet.

Though it's an idea that seems ripped from the pages of science fiction, the Hyperloop concept (or something very closely resembling it) has actually existed for centuries. In 1799, British inventor George Medhurst proposed an idea that would see goods moved pneumatically through cast-iron pipes, with carriages filled with passengers blown to their destinations. It seems even a real world Tony Stark can look to the past to inspire the innovations of the future.

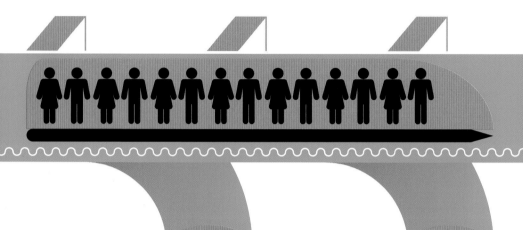

Cost-effective

Safer

HYPERFAST, OR JUST HYPERBOLE?

A successfully realized Hyperloop route could usher in massive change for those who live at either end of it. Allowing us to become near-supersonic commuters, the so-called 'commuter belt' of major cities could be expanded dramatically. This could take the strain off overcrowded metropolises where living space is costly and hard to come by. A person living in Manchester, for instance, could quite comfortably commute to London every day thanks to the speed of Hyperloop, and enjoy the lower living costs that being removed from the capital affords. Sure, any Hyperloop connection between a centre of commerce and an outlying area would likely drive up prices in the more remote location, but it would bring with it a boost to the local economy as well. It would be a relatively safe form of transportation too.

Easing congestion on roads and other forms of public transport, its controlled route would have none of the dangerous variables of a highway. And, running on electric Tesla batteries and powered largely by renewable energy, it would be safer for our health and environment to boot. Hyperloop could theoretically replace short- and medium-distance air travel, and do so with far greater energy efficiency. It would be cheaper to run and maintain than a train line, and the Hyperloop team believes operators of the systems would actually be able to turn a substantial profit – despite lower ticket prices – and without the need to rely on government subsidies.

Hyperloop is not without its detractors, and some believe it is doomed to failure. Its construction represents an intense

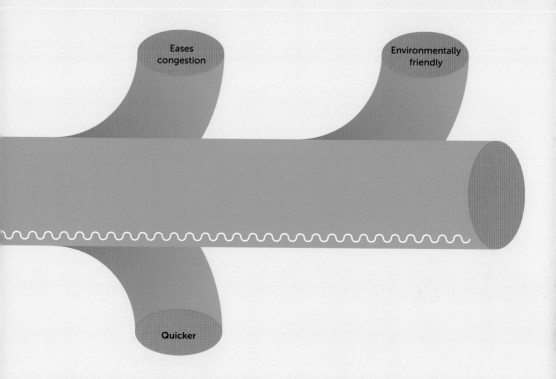

infrastructural challenge. For starters, the demanding real estate requirements seem destined to limit its application to remote locations with great swathes of empty space between potential stopping spots. Economists can't fathom how Musk intends to build such technically advanced routes at costs so much lower than traditional transportation systems either. Also, Hyperloop would represent a real target for terrorists – such a thin and exposed artery, stretching hundreds of miles, would be incredibly difficult to secure.

But the thrill of speed has always captured the imaginations of adrenaline junkies and time-strapped industry titans. Hyperloop seems both credible and environmentally aware, making it an idea worth setting in motion.

The plans for Hyperloop's construction itself offers revolutionary ideas. Taking principles from the open source computer coding development scene, Musk has essentially handed the reins of his Hyperloop idea over to the world. Holding a series of competitions, he is allowing engineering teams and design agencies to propose their solutions to the challenges of Hyperloop's ongoing development. Instead of settling on the highest bidder among contractors, those best placed to realize Musk's vision will be given an equal chance to become part of its success.

HYPERLOOP
USHER IN A
OF TRAVEL
WILL IT TAK

WILL
NEW AGE
WHERE
E US?

EXOSUITS

Homo sapiens has risen to the top of the food chain, besting the lions of the savannah and the sharks of the sea. But that's more to do with our bulging brains than our natural brawn.

From the pages of comic books to the explosive excitement of the silver screen, we're obsessed with exploring ways to make our mettle match our grey matter. And it may well turn out that those comic book fantasies have some potential usefulness in the real world too. No, we won't have to become pals with a mild-mannered alien from Krypton. But taking inspiration from Tony Stark's *Iron Man* invention may unlock abilities we've only previously dreamt of.

As futuristic as they may sound, the exosuit concept first appeared in the late 19th century, when inventor Nicholas Yagn, a Russian scientist, designed an exoskeleton-like device that used compressed gas to assist limb movements. But it has been the demands of the modern armed forces that have resulted in the most advances in the field, with billions of dollars pumped into defence firms like the US government's DARPA (Defense Advanced Research Projects Agency) to develop supercharged armours. The modern exosuit still shares many similarities with Yagn's early ideas, though battery packs and electronics are now a vital part of the equation.

Exosuits and exoskeletons look to augment the natural strengths of the human body, amplifying the advantages our bipedal frames afford us by supercharging our muscles with mechanical elements. These can be full bodysuits or modular frames to enhance particular limbs, with an interface element allowing the strong mechanical components to mirror the movements of the body parts they're partnered up with.

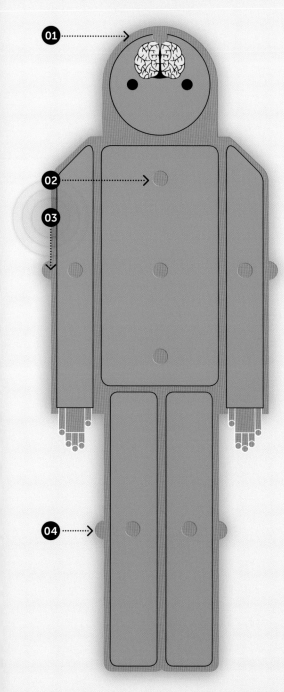

Both a full-body exosuit and exoskeleton frame will, at their most basic, make use of:

01. A mechanized frame This supports the user's limbs and joints at key points along their length.

02. Sensors Placed on the user's skin (though in time they could be interfaced directly with the brain), sensors pick up commands being sent to the muscles in the form of bio-electric signals. These must be particularly sensitive, as the voltages can be as little as 100,000 times less powerful than those that you'd find in a household battery. These sensors relay information back to a CPU.

03. CPU Receiving the information from the sensors, the CPU translates it into movement commands to be carried out by the exoskeleton. From sensor signal to CPU command, the information must be sent and processed at imperceptibly high speeds to avoid the introduction of 'lag' between a wearer's intended movements and the exoskeleton's response. A brain signal can travel almost imperceptibly at around 118m a second for an essentially instantaneous muscle response. Exoskeletons must strive to match this as best as possible.

04. Actuators These are responsible for moving the exoskeleton frame based on the CPU commands, triggering hydraulic or pneumatic power to enhance, aid or augment the capabilities of the attached limb.

SUITS FIT FOR SUPERHEROES

As with many new technologies, the main investment in exosuits and exoskeletons is coming from military organizations. They see the exosuit as a potential means of building the perfect super-soldier, but with the possibility of rolling out the powerful fantasy to divisions numbering thousands of individual troopers.

The benefits are clear to see. The average US soldier carries over 27kg worth of gear while out on operations, a weight that can be much higher for specialized soldiers such as combat medics. An exosuit could massively take the strain off manoeuvres, increasing soldiers' strength and speed, making them better prepared for combat and steadying their aim in the heat of battle.

One suit concept, known as the TALOS ('Tactical Assault Light Operator Suit'), has even experimented with liquid ceramic coatings which, when dipped with Kevlar fabric, would actually become tougher if a bullet were to strike, thanks to the way its particles would react on impact.

The metabolic cost is also worth considering – not only will wearers be able to perform these superhuman feats of endurance, but they'll be able to do so for longer before fatigue kicks in, thanks to the energy expenditure being offset by the suits. Historically, what has been funded by the military has eventually found its way into civilian life too. Imagine the rescue worker who can single-handedly lift the debris from a disaster zone, or the careworker who can lift her elderly patients without breaking a sweat during a long night shift. Any role that

requires manual labour could be made more manageable with the aid of an exoskeleton or exosuit.

If exoskeletons can revolutionize combat and the workplace, they'll have an even more profound effect on the lives of those living with disabilities. Their use will have a transformative impact on those with mobility problems, with the external aids strengthening limbs weakened by genetic and degenerative diseases. Japan's Cyberdyne Inc., with its Hybrid Assistive Limb (HAL) exoskeleton, is already using the technology for the rehabilitation of patients with weakened limbs, or those left paralyzed by a stroke. Sufferers of Parkinson's disease would also benefit from the mechanized frames, which could offset tremors to give control back to the wearer.

But it's when exoskeletons are paired with direct brain-controlled interfaces that their potential will be most fully realized. Picture a man left paralyzed by a spinal injury, preventing any signals being sent from the brain to the limbs. While the damage may be irreversible, the devastating loss of movement could potentially be reversed with the aid of a mechanized frame. Suits like the HAL already work by picking up signals from the brain using sensors placed over the muscles. If these signals could be captured at their source and sent directly to an exoskeleton frame – bypassing a damaged spinal cord – they could be used to give the wearer mobility once again. The brain will be liberated from the limitations of the flesh.

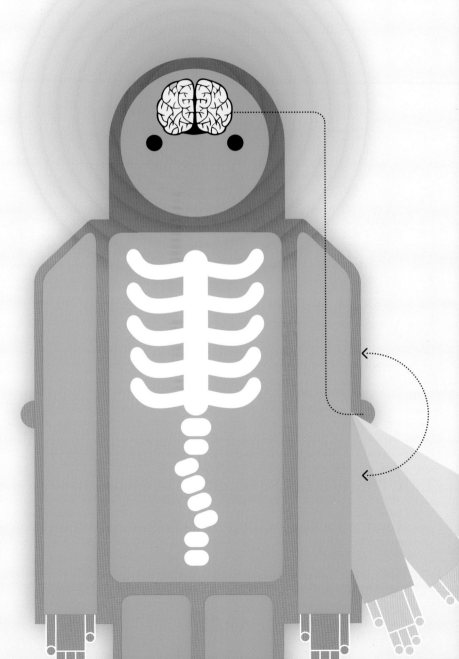

A NEW SPACE RACE

Is there life on Mars? It may be a moot question if a number of governments and private investors have their way – there will be life on Mars, once we've put ourselves up there. But actually getting people to the Red Planet? That's a challenge that only science and technology can overcome.

A new space race is underway, one in which not only long-standing government agencies like NASA, the ESA, India's ISRO and China's CNSA are taking part, but also an all-new realm of privately funded space pioneers. For many, Mars has become the number-one destination in the skies. Though its distance from Earth varies due to the planets' different orbits, the average distance of 140 million miles (225 million km) means Mars is roughly 600 times further away than the Moon. Even at its closest approach, it's 34 million miles (55 million km) away, which is about 150 times further than the Moon.

Previous missions have already sent robots to Mars (NASA's *Curiosity* rover touched down in August 2012). But can humans ever travel that incredible distance? The titans of Silicon Valley are joining the established space agencies in an attempt to find the answer. Again, it's Elon Musk's name that comes to the fore. The entrepreneur founded SpaceX in 2002, and acts as its CEO and lead designer. SpaceX may have jumped the first hurdle in a long list of serious obstacles,

creating a reusable rocket system that could lay the foundations for a new era of deep-space travel. The company's crowning achievement so far is its two-stage Falcon 9 rocket – the world's first rocket capable not only of leaving Earth's atmosphere, but also of returning to and landing safely on a floating drone barge out at sea. It paves the way for SpaceX's ambitious future BFR rockets, which are being developed with Mars as their ultimate destination.

A smooth Falcon 9 launch can be broken down into five stages:

01. Launch and separation The rocket consists of two main sections, designed to carry an attached payload out to space. At launch, the complete rocket rises to an altitude of 50 miles (80km), before the two main 'stages' separate. While the second stage continues out to space, the first stage only needs to rise to around 100 miles (160km) high before preparing for its descent.

02. Boost-back burn Using three of its engines, orientated by the rocket's guidance control system, the rocket rotates towards the awaiting drone ship upon which it will land. Guidance systems need to be incredibly accurate – the rocket is now travelling at 3,000mph (4,800km/h), and mistakes could be catastrophic.

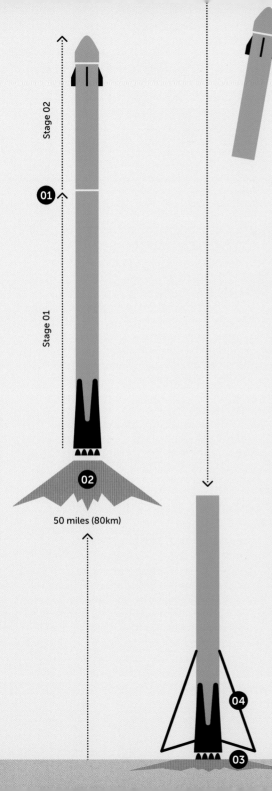

Stage 02

01

Stage 01

02

50 miles (80km)

03. Supersonic retro-propulsion burn
The central Merlin engine ignites, slowing the descent to around 550mph (885km/h). Fins extend at four points on the circumference of the rocket to slow it further and stabilize it for landing.

04. Landing burn and landing legs
One final burn takes place, slowing the rocket to a speed of just 5mph (8km/h). Four legs fold out at this stage, powered by compressed helium. They'll take the weight of the giant rocket on their carbon-fibre and aluminium frame, with a honeycomb 'crush core' absorbing impact.

05. Recovery
The rocket should now come to a relatively gentle halt upright on the unmanned drone ship at sea, where engineers can then pick up the rocket, assess its performance and vent any remaining gases before returning it to shore to be used once more. The autonomous vessel uses sensors including GPS for positioning data and four diesel-powered azimuth thrusters to maintain a steady position for the rocket, allowing leeway of 3m – even within stormy conditions.

The entire process, from launch to landing, takes less than 10 minutes.

04

03 **05**

FUTURE FRONTIERS

According to some of the greatest minds in science and engineering, from Stephen Hawking to Neil deGrasse Tyson to Elon Musk, reaching Mars isn't an option – it's a necessity. For humankind to survive into the further reaches of time, we must become a space-faring species. Successfully travelling to Mars is the first step towards that goal.

Space travel has historically been a massively expensive venture. Though that's unlikely to change in such a way as to make the Moon a potential long-weekend holiday destination, the progress made by SpaceX will help to make the pursuit more affordable. Rockets are intricate, costly machines – a Falcon 9's ability to serve multiple missions over the course of its life makes it an economic evolution of the expensively sacrificed rockets of previous decades. Reusable equipment will rapidly increase the amount of spacecraft we can send outside our atmosphere, laying the groundwork for the longer journeys that will act as stepping stones to reaching Mars.

As with any grand venture, the progress made towards a life on Mars won't exist in a vacuum. Advances in solar and nuclear energy, recycling, alternative fuel sources and food production on hostile terrain will need to be made, and the headway gained

As the fragility of our world becomes increasingly apparent, a mission to Mars will start a fire in the hearts and minds of the pioneers of tomorrow.

in these areas will have a direct impact on the quality of life back here on Earth. Rockets will get us there, but the best minds of many industries will have to pull together in order for us to spend any significant time on Mars.

It's harder to quantify, but the inspirational and morale-building value of space exploration cannot be dismissed either. The space race of the 20th century was key to inspiring children and students to excel in science, technology, engineering and mathematics (STEM) education, and any race to the Red Planet would likely spark the same enthusiasm in the 21st century. As the fragility of our world becomes increasingly apparent, a mission to Mars will start a fire in the hearts and minds of the pioneers of tomorrow – the generations for which current affairs on Earth may mean that space colonization (not just exploration) will be an obligatory part of survival.

TOOLKIT

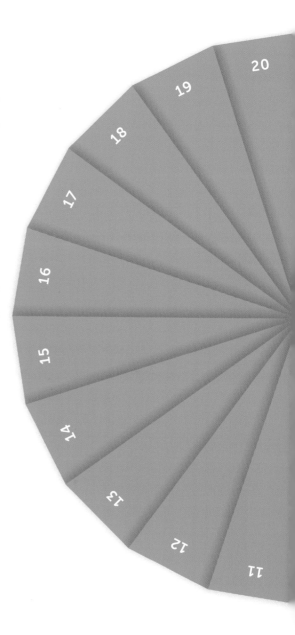

05

Driverless cars will use an array of sensors, image-recognition systems and wireless technologies to enable autonomous travel, free from the intervention of their passengers. They'll help our roads become safer and cleaner, and give freedom of movement to anyone who can access one.

06

Hyperloop will transport passengers and cargo at near-supersonic speeds, but will do so in an affordable, ecologically friendly fashion. It will transform the lives of those living at either end of its straight routes, taking the pain out of commuting and revitalizing trade and industry in even highly remote regions.

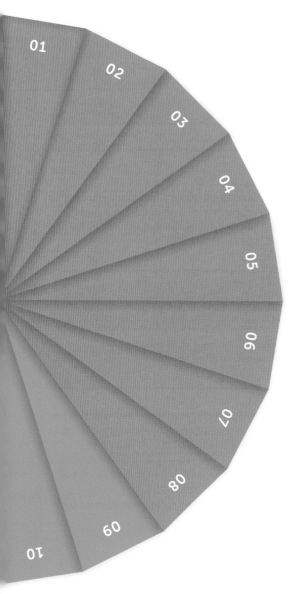

07

Exosuits and exoskeletons will strengthen our limbs, allowing us to carry out tiring, physically difficult tasks for longer, with ease. The technologies offer the possibility of giving back control to those who have lost the confident use of their limbs, from the elderly to the disabled.

08

Reaching Mars will be a great challenge, but the technology required to make the journey possible is within our grasp. Reusable rocket systems and pragmatic executives at private space exploration companies will drive down the cost of missions. Each mission will bring with it new scientific findings, inspire future generations to create their own efficient new technologies, and eventually give us the complete toolkit to embark on what may well be a one-way initial trip to the Red Planet.

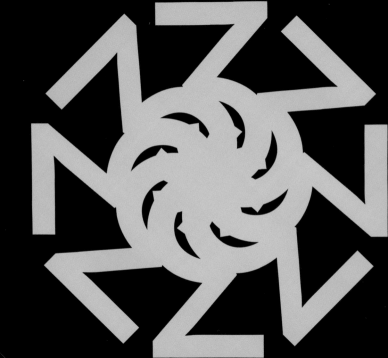

SURVIVAL

LESSONS

09   NANOROBOTS
What would be possible with a robotic surgeon the size of a pinhead?

10   THE QUANTIFIED SELF
Can personal data make us fitter, happier and healthier? Or should we be wary of a digitally determined lifestyle?

11   NUCLEAR FUSION
Looking to the stars to solve an Earth-bound problem, will we ever master nuclear fusion?

12   ASTEROID DEFENCES
Are we prepared to defend against a potential asteroid impact on Earth?

Our ability to wield technology and manipulate the laws of science will become increasingly important in the decades and centuries to come.

From fending off sabre-toothed tigers with spears to fighting off infections with antibiotics, humans have always been adept at engineering novel ways of ensuring the survival of our species. Our use of tools has allowed us to conquer the planet, climb to the top of the food chain and eradicate some of the most dangerous diseases that have ever existed.

Our ability to wield technology and manipulate the laws of science will become increasingly important in the decades and centuries to come – not only to better ourselves and improve the quality of life on Earth, but to also combat a number of existential threats that we may soon face.

Some of these threats, like that of an asteroid impact (a threat more statistically possible than you'd like to imagine), are a fluke of the infinite nature of the universe

– born beyond our realm of influence, but potentially within our power to neutralize. Others, like the growing danger of global warming, is a home-grown disaster of our own making – the 'inconvenient truth' that implores us to look into alternative green and forever-renewable energy sources.

The next few lessons will take a look at how technology and science will come together to ensure we live to fight another day in the face of some of these dangers. Some proposals envision engineering projects on a galactic scale, while others look at concept devices so small you'd need a microscope to spot them. Some ideas discuss maximizing the efficiency of our individual bodies, while others still offer an utopian glimpse at a future where energy, managed in an efficient and responsible manner, is infinitely available to all.

NANOROBOTS

In the 1966 film *Fantastic Voyage*, the members of a submarine crew shrink to a microscopic size and are inserted into the body of a sick man, in the hope that they'll be able to repair damage to his brain from the inside. Half a century later, we've not quite perfected the shrink ray, but the concept of using tiny technologies inside the body to improve health and aid recovery in sickness and injury is one that's increasingly explored – and actively developed.

Nanotechnology, and the associated robotics field of nanorobotics, looks at the ways technology can be used to influence change on matter at the nanoscale. To give some sense of perspective, a nanometre is a measurement of one billionth of a metre, or 10^{-9}m. A strand of hair is 100,000 nanometres wide, while a water molecule would measure less than a single nanometre across. So nanotechnology is looking to work accurately within a fine scale far beyond what standard microscopes can see – at a level where the very building blocks of life are at play.

There's no one defining form of nanotechnology, nor is there one leading use, but there are immediate benefits to developing nanorobotics for healthcare. If you can manipulate molecules at such a granular level, or even at a slightly larger scale, you can more easily treat and cure the root causes of disease.

Take the concept of a nanorobot designed to work within a blood vessel. Though it may lack the *Fantastic Voyage*'s human crew, the film's submarine-like craft makes a surprisingly good model for how a nanorobot could be shaped. Like a tiny torpedo, a nanorobot could travel along a blood vessel, equipped with a miniature payload of medicine or tiny tools to perform surgery with unparalleled accuracy.

Inserted into a patient inside a biodegradable pill or through a simple injection, the nanorobot could propel itself around the waterways of the circulatory system with some kind of mechanical tail, like the flagellum that allows bacteria to move around the body. Or, the robot could use a patient's own blood to move forward, generating a magnetic field to draw in conductive liquids, forcing them out through a pump and creating thrust like a water pistol.

A nanometre is a measurement of one-billionth of a metre.

A strand of hair is 100,000 nanometres wide.

DNA is 2.5 nanometres wide and a water molecule is 0.275 nanometres wide

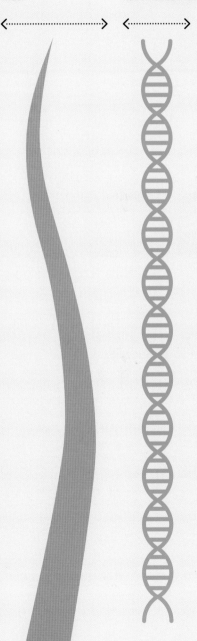

It's a great challenge finding a fuel source that's safe to use within the human body, and one that can provide enough energy to carry out the task at hand within such a small enclosure.

Nuclear power systems have been considered and dismissed, given the danger that radiation poses to human cell structure. You could link the nanorobot to an external power source, but that limits its manoeuvrability, and increases the potential for internal damage to the patient. An interesting alternative is to use the human body itself as a power source, either by inducing a chemical reaction between the nanorobot and a patient's blood to create fuel, or by fitting the nanorobot with electrodes that could interact with electrolytes found naturally in blood to create a miniature onboard battery.

SHRINKING THE SURGEON

As you can tell from the previous scenario alone, the challenges faced by engineers in nanorobotics are great, but the potential benefit to come from solving these problems is even more astounding.

Take a cancer sufferer, for instance. Depending on the form of cancer that a patient is fighting, they may currently have to contend with deeply invasive, painful surgery and a debilitating treatment programme. Both take a great physical and emotional toll on the body. A nanorobot could work within the body to cut away cancerous tissue with minute accuracy, while delivering the drugs given in a chemotherapy session direct to the source, rather than relying on the circulatory system eventually to bring them to their intended target. This has the potential to cut down the required dosage and length of a course of chemotherapy treatment considerably.

As western diets become increasingly decadent, the healthcare profession increasingly has to deal with problems created by arterial plaque. Fats like cholesterol build up within artery walls, causing the space within which blood travels through the body to narrow. A tiny nanorobot could work like a tunnel-boring machine within an artery, carefully cleaning away the fatty build-up. These procedures, rather than requiring a hospital stay, could even become outpatient treatments: you could receive a prescription from a doctor and pick up a course of pre-programmed nanorobots from your chemist, ready to be taken with a glass of water before meals.

All this is before considering the potential for nanorobots to work together. A 'swarm' of nanorobots could be deployed to carry out several tasks at once, networked together to achieve something that a single nanorobot could not – perhaps taking on protozoa or even relatively large worms.

But why stop there? The futurist Ray Kurzweil envisions a future in which we live with nanorobots inside our bodies 24 hours a day, 365 days a year, working alongside our brains to make sure we're forever at our funniest, smartest and most productive. Connecting wirelessly to cloud computing services, they could allow us to interface with the internet – and other humans – remotely at will, with just a thought. While his prediction sees this revolution beginning somewhat dubiously within the next two decades, it's easy to see this as the ultimate goal of nanorobotics.

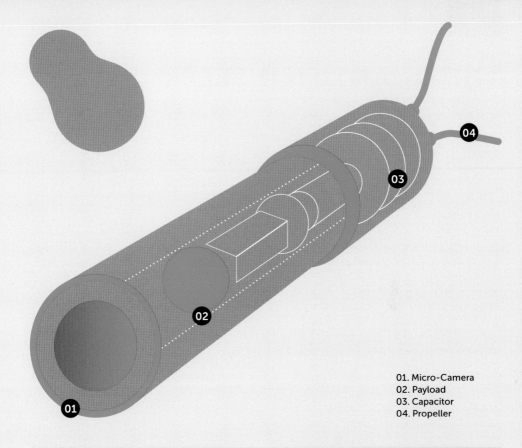

01. Micro-Camera
02. Payload
03. Capacitor
04. Propeller

Every single cell in your body contains a copy of your genetic code, known as your genome. It changes across generations, passed down by your parents, and is made up of DNA, defining your traits and characteristics, from your athleticism to your intelligence. With the entire human genome now sequenced and deciphered, we've been able to use it to help identify some of the undesirable traits too — the mutations that cause diseases.

CRISPR-Cas9 is a tool that lets us edit these sections of our DNA. It's a pair of molecules created and used by scientists to hunt down unwanted parts of DNA and cut them out, either leaving the body to then heal itself, or inserting a desired replacement DNA strand instead. Do this to reproductive or embryonic cells, and the changes are permanently passed down, creating a new inherited code.

These gene-editing techniques could be instrumental in curing hereditary diseases such as cystic fibrosis. However, they also have more controversial uses — could (and should) we be able to use these tools to create 'designer babies', or to tweak the DNA of creatures harmful to humans, wiping out entire species like malaria-carrying mosquitoes?

THE QUANTIFIED SELF

Know thyself! Since the age of the ancient Greek philosophers, we've put great importance upon being able to read and understand both our physical and inner selves. So keeping a close eye on personal metrics of all kinds isn't necessarily a new thing – for years, diligent self-study has helped athletes to push themselves to record-breaking performances, and allowed those with chronic conditions to avoid their triggers. But as processor and sensor technology has exponentially shrunk and become more powerful, so too has the technology to build a picture of a 'quantified self' exploded in popularity.

From smartphones to smartwatches, and connected rings to sensor-laden clothing, wearable and personal technologies are allowing us to track almost every aspect of our lives. Want to know your heart rate while out on a run? Strap on an Apple Watch with its photoplethysmography (PPG) reader. Struggling to get a good night's rest? Slip a Beddit Smart 3 tracker under your bed sheets and let its ballistocardiography (BCG) measure how peaceful your slumber is. Struggling to stay hydrated? Buy a Bluetooth bottle that glows when you should be taking a swig of water. There are even devices in the works that could let diabetics check up on blood-sugar levels simply through analysis of their perspiration, rather than with regular blood tests, while Google is working on smart contact lenses that can measure glucose levels when worn on the eyeball.

These are the kinds of devices that once would have been gigantic pieces of medical equipment. But they're now built and marketed to be lifestyle devices, to the point where some are being designed by high-end fashion brands.

The Fitbit Charge 2, for instance, looks like little more than a colourful digital wristwatch. But it packs in an optical heart-rate tracker that shines different wavelengths of light on your skin and notes refracted patterns in order to measure blood flow; a three-axis accelerometer measuring the movement and speed of the wearable through space to figure out your pace; an altimeter to measure altitude (or more usefully, the elevation of differing sections of a run); and a Bluetooth 4.0 radio transceiver that connects the wearable to other smart devices, where it can offload the data it has collected and receive information from the paired device's own sensors, such as GPS. The data these components generate can be shared with a whole network of apps and online fitness communities, helping to create personalized exercise plans and 'gamifying' incentives to complete those gruelling runs and reps.

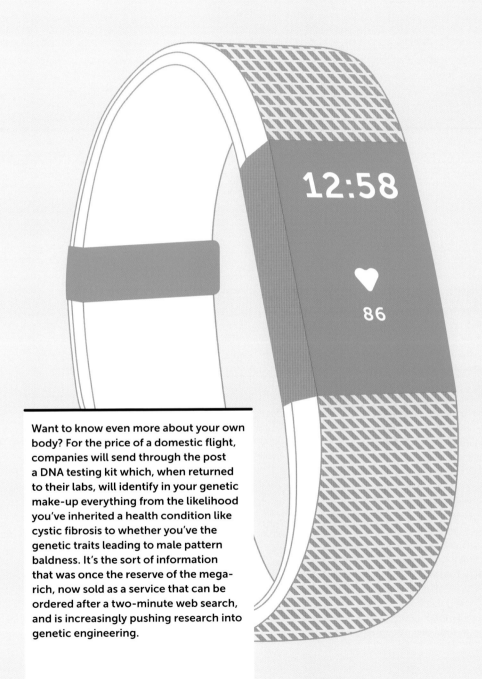

Want to know even more about your own body? For the price of a domestic flight, companies will send through the post a DNA testing kit which, when returned to their labs, will identify in your genetic make-up everything from the likelihood you've inherited a health condition like cystic fibrosis to whether you've the genetic traits leading to male pattern baldness. It's the sort of information that was once the reserve of the mega-rich, now sold as a service that can be ordered after a two-minute web search, and is increasingly pushing research into genetic engineering.

FITTER, HAPPIER, MORE PRODUCTIVE?

By tracking minutely detailed data on the make-up of our bodies and the external factors that can have an influence on us, we should, in theory, become a healthier, happier society.

We'll be able to make pre-emptive strikes on harmful conditions that would otherwise go unnoticed, take the strain off our overburdened health services by following the beneficial lifestyle advice of a series of tracking systems, and improve diagnoses across the world through the AI analysis of a more comprehensive set of medical data than history has ever known. Across the globe, health practitioners, aided by computer intelligences and an advancing understanding of our genetic make-up, will be able more quickly to spot health trends across cultures and demographics, and help combat epidemics before they ever have a chance to spread.

On a personal level, the quantified self gives individuals greater encouragement to take control of their own health and fitness. Apps and tracking devices allow for progress towards granular, tailored fitness goals to be easily monitored, as well as giving those with chronic diseases the confidence to measure their own well-being removed from their health practitioners, if required. Should anything go wrong, apps and trackers can be used to also remotely and autonomously alert a doctor to a suddenly formulating crisis, further increasing the chances of survival.

The quantified self need not be solely concerned with flesh and blood either. Using apps and software to track other metrics beyond health-focused ones can have great benefits; you could boost your savings by tracking your monthly financial outgoings, weeding out the moments when you could tighten the purse strings a little, or assess your growing abilities as you learn a new skill, such as playing an instrument or learning a language, and highlighting problem areas.

Consider also the reams of data you share through social network updates and online searches – these, when fed through algorithms, reveal great amounts about you too, leading to a quantifiable personality.

A University of Cambridge and Stanford University study found that with as few as 10 Facebook 'Likes', the social network could make more accurate predictions about your tastes and personality than work colleagues. With 150, it would outperform family members, and with 300 it could even trounce a spouse's ability to read their partner's interests. While it may help you more easily find new friends or romantic partners online, it is also the sort of data that advertisers and governments would scramble to have access to for less altruistic reasons.

This raises ethical and moral questions. Who controls and has access to this highly sensitive data, and will they keep it safe? Will those who fail to keep tabs on their personal health metrics be penalized by a data-driven society that will increasingly expect the individual to be capable of self-diagnosing and medicating? And will easy access to genetic data result in perfectly biologically engineered people where imperfections are erased at the expense of our individuality? There is great potential to be had in the coming personal data boom, but we must quantify the dangers too.

TO FIND C
ENERGY W
TAKE INSP
FROM THE

LEAN
E NEED TO
RATION
STARS.

NUCLEAR FUSION

Finding a clean, renewable energy source has become one of the modern world's most desperate goals. Positive change can begin at your doorstep, starting with taking shorter showers, and building to bigger ideas, like economical electric cars and green homes (see Lesson 04). Elon Musk's Tesla is even mass-producing a new affordable 'Solar Roof' that will store energy in a 'Powerwall' battery that can power an entire home.

However, these methods will require a unified global effort to counteract modern energy concerns effectively. So how can we provide a reliable source of power for all, affordably, cleanly and consistently?

The answer may quite literally lie in the stars. Nuclear fusion promises to provide humankind with enough energy to last us billions of years – without having a negative impact on our environment – by mimicking the makeup of our Sun. We will essentially be building a star here on Earth.

The nuclear fusion technique differs from the fission reactions found in nuclear power plants. Where fission creates great amounts of heat energy by splitting an atom (leaving dangerous nuclear waste behind), fusion looks to combine two lighter atoms into a larger one which, when perfected, will create vast amounts of clean energy. To do this, nuclear physicists look to replicate the conditions found inside the Sun, where fusion reactions are constantly happening.

There, atoms of tritium and deuterium (isotopes of hydrogen, hydrogen-3 and hydrogen-2) are each forced together under extreme combinations of pressure and heat,

resulting in the formation of a neutron and a helium isotope – along with monumental amounts of energy. It's capturing that energy here on Earth that has physicists so captivated.

Deuterium and tritium (to a lesser extent, as it must be engineered) are readily available here on Earth, with the former found in vast quantities in our seas. It's building and running the reactors that's the tricky part.

The aim is to reach the 'ignition' stage of nuclear fusion – the point where enough fusion reactions occur to enable a self-sustaining event capable of powering successive reactions, and achieving a net energy yield four times greater than nuclear fission. Though there are many experimental ways being researched with which to create a fusion reaction, the most promising is to use magnetic confinement in a vast donut-shaped structure known as a tokamak.

For the hydrogen atoms to fuse, their nuclei must come together. But as the protons in each nucleus are positively charged, they repel each other. The tokamak allows for the conditions in which the reaction can occur.

Firstly, the hydrogen gas must be turned into plasma. To do this, the tokamak uses microwaves, electricity and neutral particle beams from accelerators to heat the hydrogen to 100 million Kelvin – around six times hotter than the Sun's core, to compensate for the lower pressure being exerted on Earth. This takes vast quantities of power, making the need to hit the ignition stage so vital. As plasma, atoms are stripped of their electrons, allowing them to move

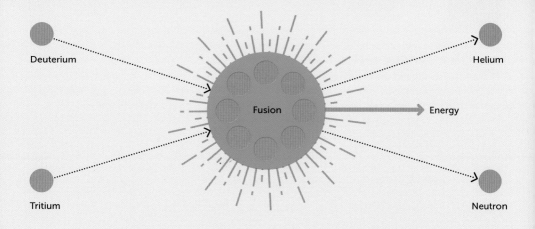

Deuterium

Tritium

Fusion

Helium

Energy

Neutron

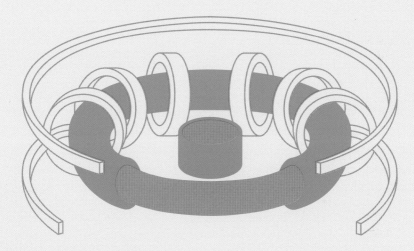

around freely. Pressure is then applied to the plasma using magnetic fields, with poloidal and toroidal coils exerting great magnetic forces throughout the donut shape, squeezing the atoms together until (ideally) the fusion reaction is sustained. With the energy produced used to heat water, a sustained reaction could theoretically infinitely power electricity-generating turbines with minimal refuelling from tokamaks anywhere on Earth.

An alternative technique is inertial confinement, in which a reactor heats a tiny pellet of deuterium and tritium fuel using focused laser or ion beams. As its surface heats to very high temperatures, the pellet implodes and massively compresses. This creates the conditions needed for fusion to occur, in turn heating the fuel and leading to the self-sustaining ignition stage.

REACHING FOR THE STARS

Despite great leaps forward in the field, nuclear fusion remains a promising pipe dream as no system yet has managed to create a fusion reaction that produces more energy than was needed to create it. The world record for fusion energy production was set back in 1997, when the UK's Joint European Torus (JET) reactor produced 16MW – enough to power a small town, but at the expense of 25MW input, and with the yield far short of the 3.86×10^{26} watts of energy that the Sun emits.

It's an incredibly complex process, but one that is slowly producing better and better results. If you were to look at a graph showing the improvement over time in the power of computer processors against the growing yields of nuclear fusion, you'd see both making gains at a comparable rate. And you'd hardly turn your nose up at what's been achieved in computing over the last half-century.

For nuclear fusion to become a viable option, it will simply take time and investment, a shift in attitudes of the oil and coal giants that hold such economic and political sway over our societies, and a step away from the radiophobia that catastrophes at Chernobyl and Fukushima have inspired.

With nuclear fusion as a concept existing since the middle of the 20th century, there's been a long-running joke that a breakthrough will eternally be 40 years away. But with talented teams and increased international co-operation in efforts to make the energy source viable, the

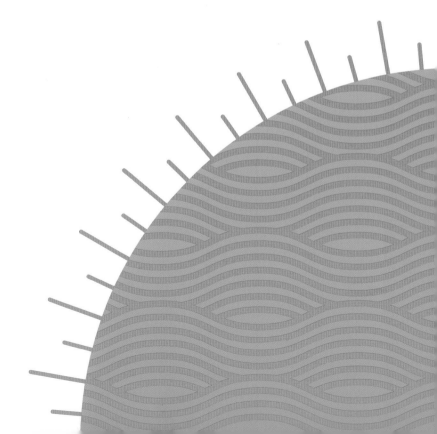

fusion idea is really starting to heat up. There are a number of sites and international teams around the world already receiving large-scale investment. There are now some 200 tokamaks around the world. To name a few facilities along with JET, there's the International Thermonuclear Experimental Reactor (ITER) project in Cadarache, France; the Mega Amp Spherical Tokamak (MAST) in the UK; and the Tokamak Fusion Test Reactor (TFTR) at Princeton in the USA, as well as the Chinese Fusion Engineering Test Reactor (CFETR). There's also the $7 billion National Ignition Facility (NIF) at the Lawrence Livermore National Laboratory (LLNL) in the US, which is focused on the inertial confinement technique.

The benefits are clear. Powered by readily-available fuel, nuclear fusion will produce far greater yields than nuclear fission and fossil fuels, without the fear of meltdown that the former brings and the pollution produced by the latter. A failed fusion reaction would see its component parts merely cool down rather than catastrophically melt down, while success would release only steam and trace radioactive elements as byproducts of the energy harvested, in quantities that would be of no cause for concern.

It may take stellar inspiration to achieve the goal, but the pioneers aiming to house a man-made star here on Earth could solve the energy and global warming crisis in one fell swoop.

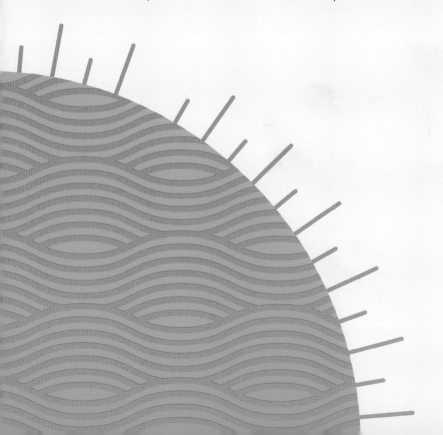

ASTEROID DEFENCES

Ever wished upon a shooting star? Then you've probably spotted a small meteoroid enter Earth's atmosphere before burning up and disintegrating ahead of impact. Breaking Earth's atmosphere is usually more structurally stressful than small objects from space can withstand. But what about larger, tougher meteorites and asteroids?

It's now largely agreed that the age of the dinosaurs was brought to an end 66 million years ago as a result of the impact of the Chicxulub asteroid. Releasing 10 billion times the energy of the 1945 Hiroshima nuclear bomb, it left a 110-mile-(180-km)-wide crater, triggering earthquakes, tsunamis and firestorms, throwing as much as 70 billion tonnes of debris into the sky and plunging the world into a two-year impact winter that blocked out the Sun. While the Earth has avoided any such similarly apocalyptic event since, the threat of impact from space objects and comets still stands. As recently as 2013, a meteorite roughly 19m in diameter struck Chelyabinsk in Russia, managing to avoid detection before impacting with the force of 450,000 tons of TNT detonating – and miraculously with no loss of life.

So, what's being done to protect us from intergalactic Armageddon?

A number of observation systems are in place to offer advanced warning of the potential dangers of near-Earth objects (NEOs), with NASA's Planetary Defense Co-ordination Office leading efforts to identify hazardous bodies. An array of worldwide ground-based telescopes help spot the NEOs of note, along with the Near-Earth Object Wide-field Infrared Survey Explorer (NEOWISE) mission – which uses a repurposed spacecraft previously tasked with making infrared scans of space to identify and classify NEOs, building a picture of their diameters and albedos (the amount of light an object reflects), and assessing their trajectories and threat level.

Identifying a NEO and its potential threat to Earth is a multi-stage process, and one that is more accurately carried out over longer stretches of time. Once a telescope finds an anomaly (usually spotted by tracking small objects moving across known, relatively stationary stars), photometric studies look at variations in brightness in the NEO over a period of time, establishing a light curve in relation to the Sun. As NEOs tend to be irregularly shaped, they will reflect light with different intensities as they travel and rotate. Once this light curve pattern starts repeating, the 'day' of the object can be identified and the amplitude of its light curve recorded, helping to define the NEO's size and shape.

To establish an orbit, multiple observations must be made, keeping in mind the gravitational pull of larger objects and planets in the solar system, and the fact that the distance from the Sun can have a bearing on speed too. As a result, it can be weeks before an orbit is confidently plotted.

A so-called 'close approach' occurs when a NEO moves within the orbit of Earth's Moon, and it's these close calls that highlight the need for planetary defence systems.

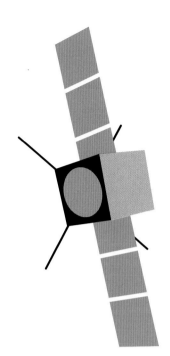

AN INTERGALACTIC GOLD RUSH

Despite scaremongering headlines, could asteroids potentially be harnessed for good?

The commercial space industry is looking into ways that asteroids can be mined for the precious metals that are increasingly scarce on Earth. Gold, silver, palladium, platinum – all manner of materials make up an asteroid composite, so prospectors have plenty of reason to set their sights on the skies.

But huge challenges face those looking to profit from space mining. Identifying which rocks will play host to rarer substances will require the development of new sensor technologies (with fuel at a premium, there's no heading to another rock if the first target for a mission proves barren), while it's very difficult to gain a steady orbit around a fast-moving, uneven object like an asteroid. Robots capable of autonomous mining of the subsurface of the space objects will be required too.

It would be history's most costly, and risky, gold rush and that's before even considering the inevitable political and economic arguments around who can lay claim to potentially vast fortunes beyond the Earth's atmosphere. But, with Earth's resources finite and depleting, for those who rise to the challenge it could prove to be history's most lucrative expedition. The *Philae* lander, which managed a nail-biting landing on a comet in August 2014 as part of the ESA's Rosetta probe mission, set a promising precedent, however.

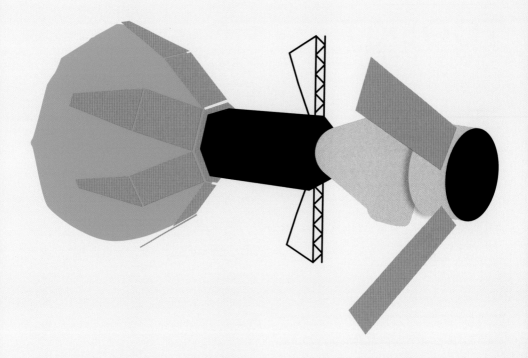

We've become adept at spotting objects larger than a kilometre in diameter. Yet if a threateningly sized object punctures the atmosphere, there is little we can do.

A series of initiatives are attempting to change this. NASA's Asteroid Redirect mission aims to send a craft with a robotic segment to a near-Earth asteroid, use a robotic arm to collect a boulder from its surface, and set it into a stable orbit around the Moon. Scientists will be able to study the sample, as well as judge what technology would be needed to redirect larger celestial bodies.

NASA is also teaming up with the European Space Agency for 2022's Asteroid Impact and Deflection Assessment (AIDA) mission. It will assess the viability of sending an asteroid off course through kinetic impact, crashing a kamikaze spacecraft into the 150m moonlet (a small body orbiting a larger space object) of the asteroid Didymos at a speed of 3.5 miles per second (6 km/s). Any noted changes in orbit can be used as a basis for calculating defences against more threatening objects.

While large asteroids do pose a potential problem, 'extinction-level' events are thought to occur only roughly every billion years. So we've probably got a little while yet to figure this one out.

TOOLKIT

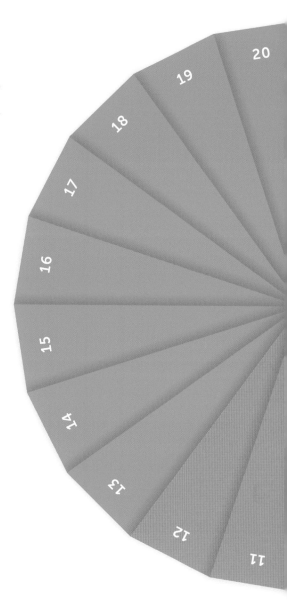

09

Nanotechnology and the related field of nanorobotics will open up whole new ways of dealing with disease and medical problems. Procedures that are today invasive and require the input of numerous health practitioners could be carried out by swarms of minuscule robots working autonomously inside our bodies.

10

Fitness-tracking devices and DNA testing, as they become ever more discreet and pervasive, will give us greater insight into how our bodies work and how best to maintain them using personalized individual recommendations, as well as exponentially increasing the amount of health data medical workers and scientists can use to invent new treatments. But the so-called 'quantified self' must be approached with caution if it's not to leave some members of society isolated and their personal data open to abuse.

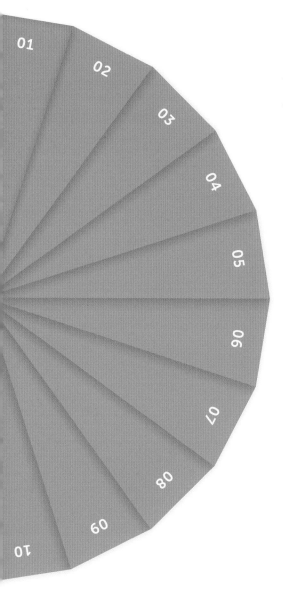

11

Nuclear fusion is a challenging, but promising, answer to Earth's problematic energy needs. If mastered, it will offer nearly limitless energy (without many of the environmentally damaging by-products of current fuel sources) by replicating the conditions inside stars here on Earth.

12

International efforts are being made to protect the planet from the danger of 'near-Earth objects' such as meteorites and asteroids. While space agencies have become adept at spotting potential threats, the coming decades will need to put unproven space defence technologies to the test in order for humanity to feel confidently protected against dangerous bodies from outside our atmosphere.

FURTHER LEARNING

READ

Engines of Creation: The Coming Era of Nanotechnology
K. Eric Drexler (Anchor, 1986)

Trackers: How Technology Is Helping Us Monitor and Improve Our Health
Richard MacManus, Rowman and Littlefield Publishers, 2015

The Quantified Self
Deborah Lupton (Polity Press, 2016)

A Piece of the Sun: the Quest for Fusion Energy
Daniel Clery
(Gerald Duckworth & Co. Ltd, 2013)

Near-Earth Objects: Finding Them Before They Find Us
David K. Yeomans
(Princeton University Press, 2016)

WATCH

An Inconvenient Truth
Al Gore's succinct and powerful documentary highlighting the need for new renewable energy resources as the threat of climate change continues to grow. Its follow- up, *An Inconvenient Sequel: Truth to Power*, is just as valuable a watch.

STUDY

Physics with Nuclear Science, University of Liverpool, UK

Electronic Engineering with Nanotechnology, University of Southampton, UK

VISIT

ITER, Manosque, France
Book a tour around the International Thermonuclear Experimental Reactor, where you can learn more about the plans for nuclear fusion around the world and get a panoramic view of the massive tokamak being constructed.

SECURITY

LESSONS

13 CYBERSECURITY
From stolen passwords to digital battlefields, how can we
defend ourselves in the age of cyber warfare?

14 BIOMETRICS
Could we one day use the unique nature of our bodies to
protect our most valued possessions?

15 BLOCKCHAIN
If we can't trust each other, can we trust the blockchain?
Looking at the digital ledger set to change the world.

16 THE AUTONOMOUS ARMY
In the combat zones of tomorrow, will autonomous robots
ever replace living soldiers?

Understanding how to operate in this vulnerable digital landscape is vital.

If technology opens new doors for us to walk through, it also allows opportunities for criminals to follow. And if technological advancements offer peace and a better standard of living for some people, that's often thanks to a trickle-down effect from the fruits of military research and warfare.

Whether on a personal or national scale, our increasingly connected lives and infrastructure face a maturing digital threat that is difficult to comprehend, let alone anticipate. Understanding how to operate safely and securely in this vulnerable new landscape is more vital than ever. All the while, the machinery of conventional warfare evolves at speed too, leading to a reliance on remote and autonomous combat units to defend us.

Researchers look to the human body to find novel new ways of protecting the things we wish to secure, while forward-thinking security experts and institutions turn to impressive new encryption methods not only to safeguard our society, but simultaneously to hold those to account who look to undermine and exploit it.

From drones in the sky to keys in our eyes, this next chapter will take a look at the ever-increasing threats our connected world faces, and the technologies being developed to defend it. It will shine a light on the techniques, hardware and machines we'll need in order to keep ourselves safe in a future world where traditional borders and barriers will lack the authority they once commanded.

CYBERSECURITY

As you sit browsing social media, an invisible war is being waged, built on an arms race in which a line of code may be as dangerous as a nuclear warhead.

The modern world has become so dependent on interconnected computer systems that the physical battlegrounds of old are increasingly being swapped for the no-man's land of cyberspace. Battles are increasingly digital and clandestine, with hackers targeting the sensitive, vulnerable digital infrastructure that underpins much of today's society.

It's no idle threat. With increasing regularity, attacks are crippling targets across the globe. Take the example of the Stuxnet virus, recognized as one of the first weapons in a new era of cyber warfare. It took advantage of zero-day vulnerabilities (digital weaknesses yet to be discovered and patched in a software package) laying dormant in Microsoft's Windows platform in order to disrupt Iran's nascent nuclear ambitions at a facility in Natanz. A type of virus known as a worm, capable of replicating itself across networked machines, it spread ferociously and specifically targeted programmable

logic controllers — unassuming computers used in industrial installations to automate a number of controls, which in this case destabilized the delicate nuclear processes. By secretly falsifying reactor readouts the Iranian scientists relied upon, the facility was rendered all but useless.

The worm's transmission? Potentially as simple as having an insider insert a USB thumb stick into a PC. Stuxnet may have been infecting systems for as long as five years before it was identified in 2010, and its reach is still being determined.

All manner of connected and critical infrastructure, services and devices are at risk from these kinds of attack. GPS satellites, land and air traffic control centres, healthcare databases, the electricity grid and nuclear stations are just some of the systems we take for granted that are in need of defence. And that's before considering the threats posed to future technologies such as networked driverless transport, automated homes and robotic workforces. Understanding how these dangers are developed and deployed is the key to protecting our infrastructure.

Stuxnet came in just one of many forms of cyber attack. It was a worm, but there are also:

01. Phishing attacks Usually distributed through malicious links in emails, these are disguised to look as though they've been sent from a trusted source, and can be used to steal user data or install malicious software on a computer.

02. Trojans Taking inspiration from Greek myth, Trojans disguise themselves as legitimate software.

03. Distributed denial of service attacks Otherwise known as DDoS attacks, these flood a network with traffic to overload and disable a system.

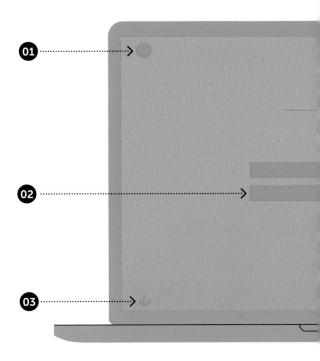

01

02

03

A CYBERPEACE TREATY

While there is plenty of investment in offensive cyber techniques, including government security agencies like the UK's GCHQ and the USA's NSA, the evolutionary nature of these attacks means that we'll inevitably be one step behind when it comes to defending against them. With culprits hard to identify and cyber criminality making a mockery of international borders, the jurisdiction of a state and its authority are left fumbling in the dark. What is a proportional response to a cyber strike? What sanctions can be brought down upon the accused should a state player be found guilty of cyber espionage? The rules of engagement have been torn up, the chess pieces scattered. Beyond chastising the phantom perpetrators, no one knows how to react to such events.

To become capable of securing individuals, institutions and states against cyber warfare, new rules must be written, and new codes of conduct established that define the limits and legalities of combat in this new realm. Could nuclear arms treaties guide us through these freshly dug trenches? War, in all its forms, brings sadness and suffering, but an internationally agreed cyber-war peace agreement, even a treaty, could limit the destruction this new form of aggression will otherwise inevitably bring.

Beyond the global theatre of cyber warfare, how can an individual limit the chances of being targeted by digital criminals? Cyber security expert Marc Goodman, author of the seminal *Future Crimes* and founder of the Future Crimes Institute, has drawn up the UPDATE protocol. This is a six-step guide to protecting yourself from up to 85% of the digital threats out to harm you.

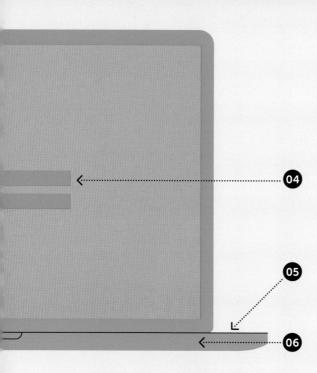

Here's a brief look at Goodman's tips:

01. Update Don't put off developer's software updates. They add new features, including patches for vulnerabilities that hackers could otherwise exploit.

02. Passwords Keep them secret, keep them safe and make them secure. Never write them down or share them, never use them for more than one device or service, and make sure they're a mixture of upper and lowercase letters, symbols and numbers. If that sounds difficult, use a password management app, like LastPass.

03. Download Only download software from trusted sources, and pay attention to the permissions (such as access to your contacts book or location) that the applications ask

for. If they look suspicious, think twice: you may be leaving yourself open to attack.

04. Admin Avoid using administrator login accounts unless absolutely necessary. Unlike standard accounts, they give deeper access to the fundamental workings of your device and its operating system. Should it be compromised, hackers will have a field day.

05. Turn off If your computer isn't on, hackers can't access it, so switch it off when not in use. Likewise, switch off connectivity options like Bluetooth and Wi-Fi when you're not using them.

06. Encrypt Should a hacker gain access to your data, they'll have a much harder time exploiting it if it's been encrypted, which scrambles the information.

BIOMETRICS

We've all got things we want to protect. Whether physical (like the contents of our homes) or digital (like the information stored on our Facebook pages), we're constantly locking things away, hoping to keep prying eyes and sneaking fingers away from our valuables.

It doesn't take Houdini to pick a physical lock, and with the right software and know-how a digital password can be hacked in minutes. The latter is particularly vulnerable – as we encounter more digital devices and services, we're required to remember ever more passwords and usernames. This leads to the formation of bad habits for the sake of convenience – creating easy-to-remember details or regurgitating the same passwords for multiple services (leaving each equally open to attack if one is compromised).

Biometric systems are poised to become the hassle-free, secure answer to modern-day security woes. Rather than relying on something we have (like a keycard) or something we know (like a pin number), biometric security is built around something we *are*.

It takes a unique physiological identifier from a user – say your fingerprint or vocal sound wave – and uses that to identify and authenticate you as the right person to access or view an item. Your body becomes a biological key. There are already plenty of locks it can be used to work with – laptops and smartphones come with fingerprint scanners; digital passports are routinely paired with facial-recognition cameras; and some banking institutions allow access to an account authenticated via voice recognition.

These one-of-a-kind organic pin codes are with us 24 hours a day, seven days a week, ready to be used as needed – whether to unlock digital data, or in conjunction with physical barriers, to let us access locked-down locations and storage in the real world.

The most accurate known biometric security method is an iris scan. It allows you to unlock something with just a stare from your eyeballs. The iris is an inherently secure

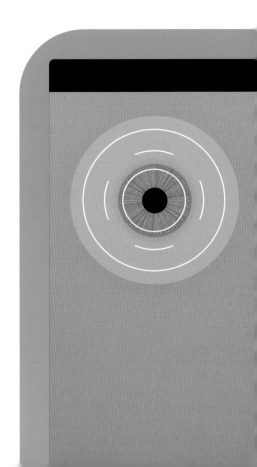

body part. Each person's iris colour and pattern is complex and unique and, unlike the fingerprint, it doesn't wear away over time as it is naturally protected by the eyelid.

01. First, your eye must be 'enrolled' to a database and associated with your personal identity. The iris is digitally photographed, in both ordinary light and the longer, invisible infrared wavelength, which helps identify details in darker eyes.

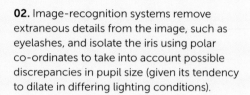

02. Image-recognition systems remove extraneous details from the image, such as eyelashes, and isolate the iris using polar co-ordinates to take into account possible discrepancies in pupil size (given its tendency to dilate in differing lighting conditions).

03. Hundreds of individual identifiers in the eye are then isolated using bandpass filters, which can note differing brightness levels in areas of an image, unique to each iris.

04. The images are converted into digital (encrypted) data, and stored on a secure database along with your personal details.

05. A scanner linked to the database remotely from anywhere in the world can take images of your eye, and compare it to the known data it has stored. If it finds a match (and can verify you're who you claim to be), the metaphorical door is unlocked and opened.

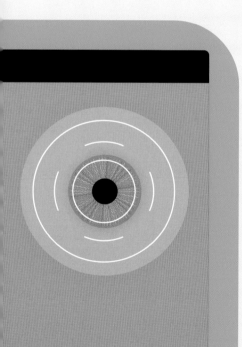

Biometric security methods are at their strongest when part of a two-factor authentication system. An iris scan is doubly effective if it works with an alphanumeric password. This is made more secure when a physical device is present; a pin code can be sent to your phone when you try to unlock a compromised account, in the statistically safe hope that you've not lost both password and phone.

PERSONALIZING PROTECTION

Biometric security will help our digital and physical lives become more secure, protected against an increasingly technologically savvy criminal mob. By harnessing what makes each of us unique, we'll be able to protect our data and property from those that would take advantage of honest and hardworking members of society.

Even the most forgetful of people will find that their valuables become better secured – biometric identifiers can't be lost and can't be forgotten. And this will have knock-on financial benefits for all – IT administrators will use fewer resources securing compromised accounts, premises managers won't need to charge for lost keycards, personal finances will be secured thanks to lowered instances of banking fraud, and police funds could be spent elsewhere rather than chasing down on- and-offline data crooks.

But, like all defence systems, biometric security is not invulnerable to attack. The fear, of course, is that these measures will become the hackers' prey of tomorrow, a potential quarry that could have lifelong ramifications for the person being targeted.

As our bodily scans are turned into digital data, can we trust the custodians of this information to defend its integrity and maintain its privacy as we would ourselves? When biometric log-ins become the default security system, what happens to someone who has already been hacked, with their iris scans and fingerprints circulated through the digital underworld? You can change an alphanumeric password or a metal key, but your voice, eyes and fingertips are with you forever. It's been known for poorly implemented facial-recognition systems to be circumvented with a high-resolution photograph, and for low-rent fingerprint scanners to be fooled by pressed modelling clay. If their more advanced counterparts are compromised in the future, your entire connected life could become permanently thrown into disarray.

Biometrics need not necessarily stop at ways with which we can lock or unlock our lives' most precious aspects either. It can also be used as a means of surveillance. There's growing interest in the so-called area of 'behaviourmetrics', the ability to identify a person not just through their bodily traits, but by what they do too. Your walking gait, the beat of your heart, the speed and pressure with which you attack the keys on your laptop – with the right monitoring sensors and software, these can all be used alongside other data correlations to distinguish you from the pack. Whether used to target billboard advertising specifically to your tastes as you pass by, or to discern aggressive intent in a fanatic's social media rants, your body and its actions will let you become tracked and analyzed in ways previously unheard of.

As with all sensitive data and materials, biometrics will need to be treated with care and respect. As their use becomes ubiquitous, we will need to work not to take them for granted. But with a diligent approach to our personal security, and a knowledge of the threats it faces, biometric systems will be a powerful additional shield in the fight against future fraudsters.

BLOCKCHAIN
REVOLUTION
SECURITY S
WE TAKE FO

COULD
IZE THE
YSTEMS
R GRANTED.

BLOCKCHAIN

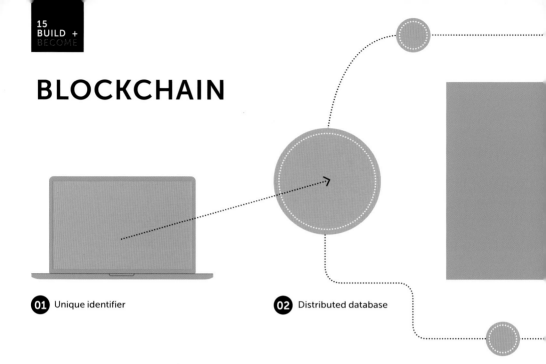

01 Unique identifier

02 Distributed database

You're on the stand, facing down a judge and jury for a crime you know you didn't commit. But if the gavel-wielding official had a decent internet connection and access to a blockchain log of your life, the judge would already know well in advance of your trial whether or not you've been a saintly, law-abiding citizen. And, with the blockchain, it wouldn't be solely those in appointed positions of authority keeping track of your actions, but a whole network logging irrefutable, immutable proof of your good deeds and wrongdoings.

Private, peer-to-peer and permanent, blockchain is a decentralized digital ledger system, building upon the principles of the databases first built at IBM in the 1970s, but valuing transparency, accuracy and incorruptibility above all else. It has the potential to transform the power balance of any relationship of a transactional nature – be that between you and your bank,

a government agency or even your partner – by keeping a historic log of a series of events that cannot be faked or tampered with, secured by a networked swarm of witnesses.

Here's how a blockchain works:

01. A potential piece of information (anything from a card transaction to the date an apple was picked from an orchard), is assigned a unique identifier, made of a 30-plus character alphanumeric code. This nugget of information is a 'block' that will join a 'chain' of other blocks representing specific data.

02. The information is placed in a distributed database that every party related to this blockchain has access to. Anyone can verify the records of any transaction without a middleman, as all parties can view the data on all previously logged events. With the entire dataset accessible to all users, no single person accessing the blockchain

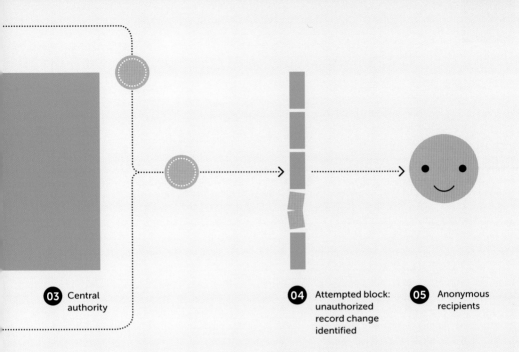

03 Central authority

04 Attempted block: unauthorized record change identified

05 Anonymous recipients

is able to own (or withhold) information. All parties must approve an exchange before it can be recorded.

03. Peer-to-peer transmission of data keeps all records up to date, and bypasses a central authority. Instead, each individual tapping into the blockchain acts as a node passing on the information to several other parties.

04. Thanks to the unique identifier each block in the chain has, it's impossible to reverse or alter records in a blockchain. Each block in a chain is assigned its unique alphanumeric code based on the code of the previous blocks in the chain. As blocks can only be added in linear, chronological order, the moment one block in the series is removed or altered, the following block identifiers would fail to match up with what's gone before, highlighting the moment the metaphorical links in the chain were broken.

05. Though the value of each transaction and block in the chain is visible to all users of the system, the users themselves can choose to remain anonymous if they wish. A blockchain address is all that's needed to allow a transaction to take place.

The most high-profile example of the blockchain in action today is the international cryptocurrency Bitcoin. A private and direct way to make payments between users, its use of the blockchain caused an algorithmic gold rush. 'Bitcoin miners' were rewarded with the currency for using their computers to process some of the complex mathematical computing problems generated by blockchain transactions.

Early Bitcoin prospectors now stand the chance of becoming very rich indeed. November 2017 saw a single Bitcoin worth a then-record high of $11,000 (£8,191). This is just the beginning for the blockchain.

TRUST IN THE AGE OF IRREFUTABLE EVIDENCE

As the blockchain is increasingly used, we – as nations, institutions, businesses and individuals – will become more accountable. Accountability breeds responsibility, and responsibility leads to measured actions and the understanding of consequences. Whether in a global bank or a local police force, an implemented blockchain will shine a light on any indiscretions, empowering us to question those institutions history has taught us to trust unquestioningly.

Blockchain's strengths may sound impersonal, holding 'the man' to account. But the world's most vulnerable individuals could be protected by its use too. Blood diamonds and the awful conditions in which they are mined could be better policed, with the provenance of the gemstone tracked from source to jewellery seller, ensuring the working environment of those in the mines. Human trafficking rings could also be fought – 1.5 billion people around the globe have no formal identification documents, leaving them at risk of exploitation. Blockchain could be the framework to protect this invisible society, calling into question the movement of people under duress, the distribution of profits and identifying stages in a manufacturing process where a human labour cost is being obfuscated.

On a personal, everyday scale, the benefits of widespread blockchain adoption would quickly become apparent too, particularly where loosely managed paper trails still cause problems. Banking fraud and economic crime could be reduced significantly, protecting your savings and investments. Medical records could be set in stone, along with the history of any treatments for a patient – and of those who prescribed them. Insurance claims could be massively streamlined, with irrefutable evidence of a claim backed up by the definitive nature of blockchain data.

The irony, of course, is that the blockchain system, essentially built to circumvent the elephantine institutions of old and the prying eyes of governments, can just as easily be co-opted by the very forces its early evangelists are looking to devalue – it is a technology that enables the watching and monitoring of people. While accountability is welcome, the permanence of the blockchain, at least in some of its potential applications, brings with it great moral questions. What room is there for interpretation in a dataset of irrefutable events, and what will happen to the value of trust?

The benefits of widespread blockchain adoption would quickly become apparent where loosely managed paper trails still cause problems. Banking fraud and economic crime could be reduced significantly, protecting your savings and investments.

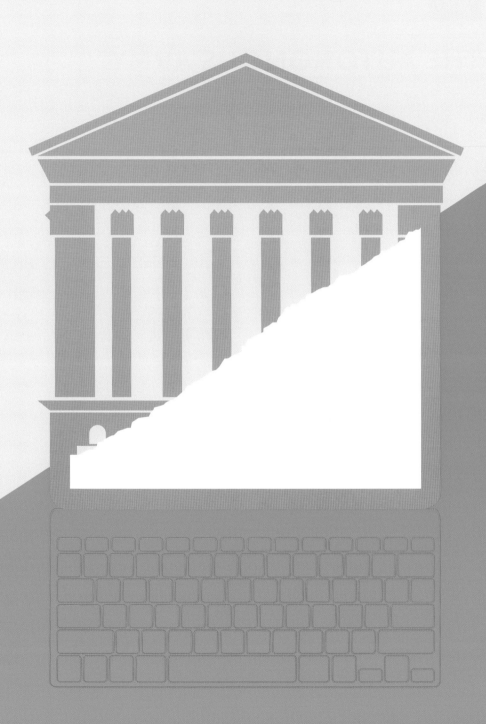

THE AUTONOMOUS ARMY

Is it a bird? Is it a plane? Look to the skies in the near future and chances are it will just as likely be a drone flying up above your head.

Unmanned aircraft of various sizes with advancing autonomous capabilities are increasingly used in civilian airspace. Usually equipped with high-resolution cameras, they've been employed by Hollywood, transported medical supplies in Rwanda and remotely piloted by rescue teams searching for survivors in unreachable disaster zones. You may have found a recreational drone under your Christmas tree, while blue-sky thinking Amazon is even imagining autonomous drone fleets housed in airships, ready to parachute parcels onto your doorstep.

Drones have their roots in military hardware. Evolving from the tactical secrecy of the high-altitude U2 spy planes, unmanned drones were initially used primarily for surveillance. In 2002, just over a year after the 9/11 World Trade Center attack sparked that decade's 'War on Terror', the CIA ordered its first drone-launched Hellfire missile attack (its target Osama Bin Laden). The world's military has never looked back since.

Of its increasingly varied drone army, the US military's MQ-1 Predator and MQ-9 Reaper UAVs (unmanned aerial vehicles) are perhaps the most widely used, and deadly. Though regularly armed, the Predator is more often used for intelligence, surveillance and reconnaissance operations.

But what about forces on the ground? Will they too be roboticized? Some already are. The Foster-Miller Talon, is one of many remotely operated pint-sized tanks in operation, alongside the MUTT (multi-utility tactical transport).

The animal kingdom is also inspiring the development of skilful robots. Boston Dynamics' Atlas is a bipedal robot with a high strength-to-weight ratio. The company also designed Big Dog, an incredibly tough machine that carries 150kg loads on its four legs; the WildCat, a cheetah-like bot that can reach speeds of 20mph (32 km/h); and the SandFlea, which is capable of jumping 10m into the air.

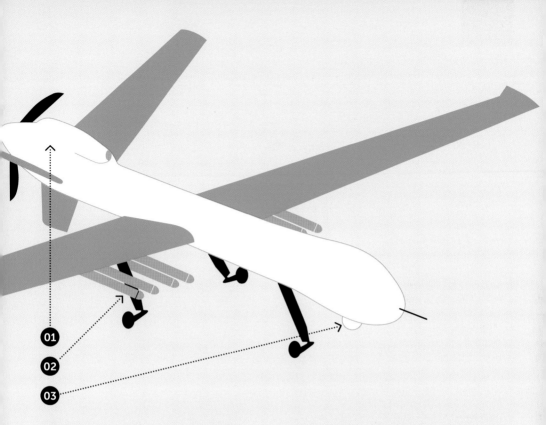

At 11m long, the Reaper (illustrated) has a 20m wingspan and is capable of flying to a range of 1,865 miles (3,000km) at 230mph (370km/h), remaining airborne for as long as 42 hours without refuelling. With a potential total payload of 1.7 tonnes of ordnance, each Reaper is remotely operated, with pilots hundreds of miles away carrying out dangerous missions from a position of safety.

01. Synthetic aperture radar (SAR) This is a form of radar antenna used to create three-dimensional images of objects by sending out pulses of radio waves and recording the returned echo. It supports the use of the GBU-38 JDAM (Joint Direct Attack Munition), a so-called 'smart bomb' conversion kit that can be fitted to the munitions that work in tandem with the SAR, giving them an integrated inertial guidance system paired with a GPS receiver for improved long-range accuracy.

02. Munitions These include the 230kg laser-guided Guided Bomb Unit-12 Paveway II, as many as four Air-to-Ground Missile (AGM)-114 Hellfire missiles and the infrared-homing AIM-9 Sidewinder.

03. Multi-Spectral Targeting Systems (MTS) Allowing remote pilots to see the battlefield, these offer target tracking, acquisition and laser guidance. With adjustable fields of view and zoom levels in the visual and infrared light spectrums they offer actionable video feeds from miles above the ground.

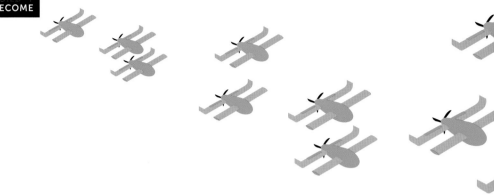

MAN VS MACHINE

The direction that drone development is ultimately moving towards is an AI-driven future, one in which machines can be commanded to plan and execute attacks on targets without human intervention.

The fruits of the research are already being seen. Small Perdix drones are just one model being designed for and tested in large co-ordinated group manoeuvres. Inspired by insect swarm behaviours, they use a networked, shared artificial intelligence system to work in formation – a hive-mind approach to surveilling or overwhelming a target. They could be used for reconnaissance, to jam enemy communications, act as radar decoys or ultimately carry weaponized loads.

The autonomous army will be a critical force for any military capable of wielding it. Reducing the amount of danger allied human combatants face, conflicts could resolve quickly and decisively, while surveillance campaigns using increasingly impressive imaging systems will act as a deterrent against any enemy with aggressive intentions.

However, the world must tread carefully around the development of self-sufficient soldiers. The key question is, where does this drive for automation stop? Right now, autonomous combatants require a kill command from a human operator before taking down a target. There are obvious dangers. Consider a scenario in which a human-controlled robot soldier or drone comes up against a fully autonomous killing machine that is not required to wait for permission to pull the trigger. A soldier that cannot react faster than its enemy is operationally useless, and a split-second's moral validation will be perceived as a weakness if it means a thoughtful combatant, human or otherwise, is a fatality certainty.

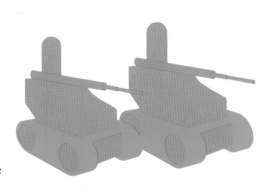

Small Perdix drones are just one model being designed for and tested in large co-ordinated group manoeuvres. Inspired by insect swarm behaviours, they use a networked, shared artificial intelligence system to work in formation — a hive mind approach to surveilling or overwhelming a target.

However, the moment of human recognition is key to working out how to create a results-focused synthetic soldier with a moral code. It is a vital challenge that must be met before further developments can be made. As such, tech and science luminaries, including Steve Wozniak, Demis Hassabis, Jaan Tallinn, Elon Musk and Stephen Hawking, were among more than 1,000 signatories of a letter presented at the 24th International Joint Conference on Artificial Intelligence in 2015 warning of the existential threat of AI-powered combatants.

An autonomous army may lessen the amount of danger and terrors a human troop has to face, but it also distances us from the horrors we create when engaging in armed combat. We fear 'The Terminator' because it has no compassion, remorse or pity — and we will have to ask what we have allowed ourselves to become if we create a military culture in which we're happy to order the execution of an attack remotely while shying away from the deathly consequences that we have wrought.

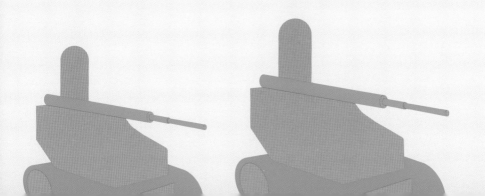

TOOLKIT

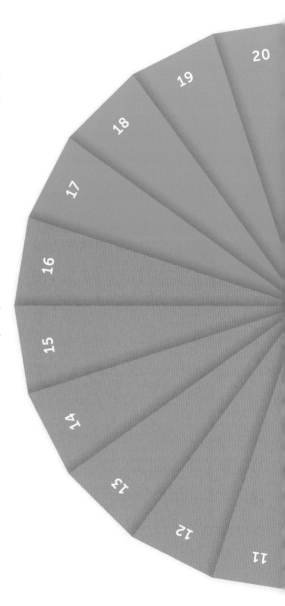

13

A connected world has its benefits, but those connections leave us vulnerable to all-new forms of attack. Unless we take the proper precautions, either by following more stringent cyber security methods or by campaigning for international cyber-war treaties, everything from our power grid to our bank accounts – and even connected homeware devices like fridges and smart speakers – could become prey for hackers.

14

In the future, passwords and iron keys will go the way of the dodo in favour of biometric security systems that make use of the uniqueness of our individual bodies to protect our valuables. Though they're not invulnerable to attack, security methods like fingerprint and eye scans add a complex level of defence to deter would-be digital intruders. However, the gatekeepers of our biometric data must treat our details sensitively lest our privacy be abused, and we must remain vigilant to ensure that biometric systems do not pave the way for invasive means of surveillance.

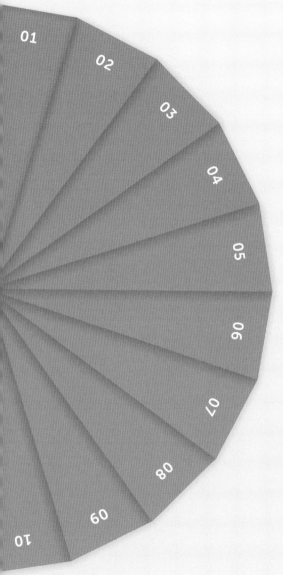

15

Blockchain's peer-to-peer and distributed design has the potential to prevent fraud and corruption from taking root wherever interactions of a transactional nature are present, as well as introducing novel new forms of finance and currency. In the wrong hands, however, it can be abused, with the power to offer governments and institutions unprecedented means to track the public.

16

When physical conflict becomes unavoidable, we'll increasingly turn to drones, robots and other machines to take on our enemies while keeping our flesh-and-blood troops out of the firing line. In the future, the autonomous abilities of these silicon soldiers will expand dramatically, and we will have to review the moral and existential implications of letting machines choose which humans live or die.

FURTHER LEARNING

READ

Future Crimes
Marc Goodman (Doubleday Books, 2015)

We Are Anonymous
Parmy Olson (William Heinemann, 2013)

Digital Gold: The Untold Story of Bitcoin
Nathaniel Popper (Penguin, 2016)

Kill Chain: Rise of the High-Tech Assassins
Andrew Cockburn (Picador USA, 2016)

**Dark Territory:
The Secret History of Cyber War**
Fred Kaplan (Simon & Schuster, 2016)

WATCH

**The Internet's Own Boy: the Story of
Aaron Swartz**
A documentary on the life of programming
prodigy and information activist Aaron
Swartz, who committed suicide aged just 26.

STUDY

Cyber security
The University of San Antonio, Texas, USA

Forensic computing and security
UWE Bristol, UK

VISIT

Bletchley Park, Milton Keynes, UK
The site where computer pioneer Alan Turing
broke the Nazis' Second World War Enigma
code, in what was arguably one of the first
acts of cyber espionage.

TRANSCENDENCE

LESSONS

17 QUANTUM COMPUTING
Can we go beyond the bytes and bits of today's machines and unlock new computational powers?

18 TERRAFORMING
We may one day be able to get to Mars — but how will we make it home?

19 BIONIC IMPLANTS
Whether with a zoom-powered eyeball or a crushing fist, will tech embedded in our bodies let us leapfrog evolution?

20 TRANSHUMANISM
Can we use tech and science to halt the process of ageing and cheat death itself?

We may soon find ourselves on the cusp of transcending the limitations of our current computer systems, our planet, our bodies and even our minds — leapfrogging the ideas of evolution that Charles Darwin laid down more than 150 years ago.

As we've explored over the preceding pages, technology and scientific progress present the opportunity to radically redefine our world and our place in it, revolutionizing industries and the way we live our lives. We have the power to influence the direction in which these advances take our society — so where would we like our ultimate destination to be, both literally and philosophically?

Should we safely and wisely navigate the possibilities that advanced technologies give to us, we may soon find ourselves on the cusp of transcending the limitations of our current computer systems, our planet, our bodies and even our minds — leapfrogging the ideas of evolution that Charles Darwin laid down more than 150 years ago.

This chapter will discuss those most future-gazing of technologies and concepts. It will look at the quantum physics that could hold the key to an explosion in new computing capabilities, and what that will allow us to achieve; the science that could potentially allow us to tame harsh and distant worlds, ready for the first interplanetary human colonists; the body modifications that will bring man and machine closer than ever before. And, finally, it will discuss the growing transhumanist movement, and why more and more people see brain-uploading technology as the key to eternal life.

QUANTUM COMPUTING

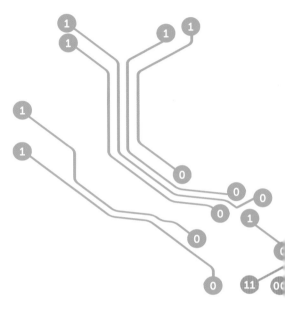

Moore's Law, named after its author and Intel co-founder Gordon Moore, dictates that the power of computer processing chips will double every 12 months. The theory was established in 1965, but was revised in 1975 to shift that expectation in processing growth to every two years. It's a theory that has more or less held firm ever since, with the number of transistors squeezed onto tiny chips growing at an incredible rate.

However, while it's now possible to pack billions of transistors onto a chip the size of a penny, the laws of physics dictate that we'll eventually run up against the limitations of working at an atomic scale with conventional computing methods. And that's where quantum computing comes in.

Based on findings from the field of quantum theory, it looks at how we can build supercomputers that work by manipulating atoms and subatomic particles using new and exciting techniques. At this infinitesimally small size, the laws of physics work in fundamentally different ways. Those working in quantum computing believe that by exploiting these differences, we'll be able to create computer systems exponentially more powerful than those that dominate the world today.

01. Whereas conventional computers at their most basic work by storing information in bits, a quantum computer would use quantum bits (represented by quantum particles like photons and electrons), otherwise known as 'qubits'.

02. While the traditional bit can only store values as 'on' or 'off' (often written as the '1's or '0's that chain together creating the binary code that underlies all current computer processing instructions), a qubit is capable of existing in a superposition – as a 1 and 0 at the same time. With every additional qubit,

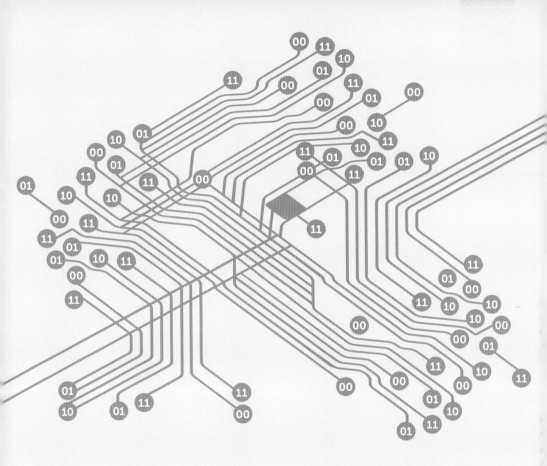

the possibilities grow exponentially; add a second qubit and you'd get not only the 00, 01, 10 and 11 states, for instance, but also all of those positions at the same time. In mathematical terms, it's the same as saying that with 'n' qubits, you can represent 2^n states simultaneously.

03. Multiple qubits can encounter a state known as 'quantum entanglement' too. It sounds messy, but it's actually beneficial to quantum processing – entangled particles become interdependent, letting them act as a single system. As a result, where

a conventional computer works through processes in sequence, a quantum computer, using entangled qubits' superpositions, could see the qubits work in parallel, attempting untold permutations on a problem's possible answer at once.

This could result in computers millions of times more powerful than those we're currently familiar with. At least, that is, in the areas in which quantum computers are expected to excel.

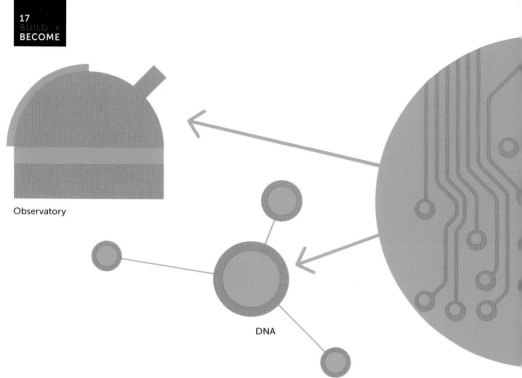

Observatory

DNA

COMPUTING THE UNKNOWN

The heart-racing thing about quantum computing is that it's an area of both physics and computing that's so far out on the verges of what is theoretically possible with a computer that we're yet to really know exactly what it can be used for – and what it will allow mankind to become.

What we are confident about, however, is that quantum computers will be perfectly suited to dealing with calculations where testing uncountable potential solutions to a problem could lead to a breakthrough. One such example is factorization – uncovering two unknown prime numbers that can be multiplied together to find a known third number. This single example is important as many of the encryption techniques used to protect our digital data – from banking details to mobile messaging apps – rely on the fact that conventional computers

struggle to carry out this sort of task. Where it would take years for current computers even to attempt to crack these systems, a quantum computer could potentially achieve a result in hours. Cryptography as we know it would be rendered obsolete.

While that's a minefield that will eventually tentatively have to be walked, the unrealized capabilities of quantum computing have a massive capacity to do good too – overlapping with many of the ideas and areas already discussed in this book.

The image-recognition systems of driverless cars could get a boost through quantum computing access, not to mention allowing autonomous vehicles to navigate the most efficient route to a destination through analyzing reams of traffic data. DNA sequencing and amino acid mapping could be carried out at superfast speeds, allowing

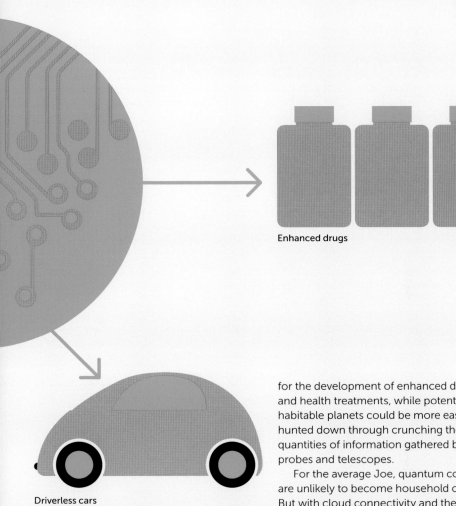

Enhanced drugs

Driverless cars

For the average Joe, quantum computers are unlikely to become household objects. But with cloud connectivity and the internet of Things, we may be able to tap into their resources remotely for specific tasks through other devices.

for the development of enhanced drugs and health treatments, while potentially habitable planets could be more easily hunted down through crunching the vast quantities of information gathered by space probes and telescopes.

For the average Joe, quantum computers are unlikely to become household objects. But with cloud connectivity and the internet of Things, we may be able to tap into their resources remotely for specific tasks through other devices. An AI-powered helper, for instance, would become dramatically more useful if it could access quantum computing powers to interpret your needs better.

We're still some way off from a world in which quantum computing becomes a mastered, stable norm. But when the age of the qubit is ushered in, it will be the beginning of a new era of discovery.

TERRAFORMING

Evolution has allowed humans to grow into the most technologically advanced species in the known universe. But our evolutionary prowess comes with a large caveat – we are intrinsically linked to planet Earth.

Our lungs only work in an oxygen-and-nitrogen-rich atmosphere, our bones can only withstand the pressures of a specific window of gravitational forces, and we draw our energy from food that grows in nutritionally rich soil. How could we ever survive anywhere else?

Terraforming is one theoretical, multi-faceted possible solution to our deep-space quandary. It explores the ways in which we can use our technological achievements and scientific knowledge to adapt hostile worlds to our needs, with Mars (being the nearest planet with a comparable make-up to Earth, sitting in our Solar System's so-called habitable 'Goldilocks zone') theoretically the most feasible candidate for adapting.

The first challenge would be to alter the make-up of the Red Planet's atmosphere. Compared to Earth, its atmosphere is 100 times thinner and made up of 95% carbon dioxide. It's much colder on Mars too, with temperatures averaging out at around -60 degrees Celsius. To compound matters further, Mars has a far less significant magnetosphere than Earth, meaning there would be no protection against dangerous, charged radioactive particles bombarding the planet from space.

On Earth, we consider global warming a harmful thing. On Mars, however, we could use it to our advantage, creating an intentional, accelerated temperature increase across the planet. If enough carbon dioxide could be captured or methane mined from the surface of Mars, this would be – again, theoretically – a reasonably easy (if slow) process.

A more dramatic method would be to redirect comets into Mars's surface. Rich in ice and ammonia, the comets would release water vapour into the atmosphere, while the ammonia could be converted into the nitrogen Mars severely lacks. An equally bombastic approach would be to drop thermonuclear weapons over Mars's frozen poles, turning them to water and in turn also helping to build a thicker atmosphere. The fallout associated with nuclear detonations would be swept away quickly, as the thin atmosphere would allow it to escape. Treat the soil, plant some vegetation, and suddenly things start beginning to look a lot more like home.

The planet comes with a catch tied to its lack of an Earth-like magnetosphere. Any efforts to change the planet would either need to be defended against or made to outpace the damage caused by radiation and solar winds, as the magnetic field would not be able to hold the engineered atmosphere steady against the solar erosion.

The southern hemisphere of Mars has a stronger magnetic field, making it a better first stop for localized efforts, but it is still nowhere near strong enough to guarantee success. Liquefying the planet's dormant outer core (again with nuclear weapons) could be one way of attempting to generate the dynamo-like, current-producing processes witnessed on Earth. But even then, the effects could be futilely temporary – with one-tenth the mass of our planet and only a fraction of its gravitational pull, the atmosphere could simply become lost in space.

INTERPLANETARY IMMIGRATION

While overhauling a planet to make a new garden world is the space-faring utopian ideal, these are, at best, ideas that will take generations to develop. It's more realistic that we'll set up long-term outposts on distant planets in our quest to colonize the cosmos. Believe it or not, these structures are already being designed, and finding a solution to an environment as inhospitable as Mars could teach us about new ways of living on Earth.

The long journeys, limited resources and harsh climates of off-Earth living bring their own challenges. With little immediate hope of a ticket home for the first Martian pioneers, any habitats need to be lightweight, reliable and incredibly tough.

One potential concept for sustained (and sustainable) Red Planet habitation is the Mars Ice Home, designed by researchers at NASA's Langley Research Center.

The inflatable living space, shaped like a torus, would be lightweight enough to transport on the long trip to Mars — and light enough for simple robotic systems to construct ahead of a human crew's arrival. Looking a little like an igloo, it would share another trait with the Arctic abode — its walls would be inflated with ice. If, as commonly thought, frozen water will be found on Mars, it could be extracted to fill the structure.

Using ice in the construction would be beneficial for a number of reasons beyond its assumed abundance on Mars, notably the fact its hydrogen-rich nature would make it a natural shield against the high-energy radiation that would otherwise bear down on a crew. It's also translucent, letting astronauts enjoy a day and night cycle, which is obviously preferable to other enclosed and underground designs that have also been proposed. According to the NASA design team, the Mars Ice Home could be put together in the 400 days preceding a crew's arrival, and could include not only crew quarters, but also a library, lab and sunlight-capturing greenhouse. A pocket of carbon dioxide, also readily available on Mars, would act as an insulator.

01. Airlock
02. Soft hatch
03. Greenhouse
04. Crew quarters
05. Laboratory
06. Gas insulation pockets
07. Ice chambers

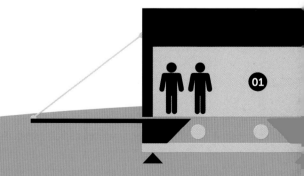

Terraforming need not initially take place on a planetary scale either. Depending on the techniques employed, these smaller settlements could be established with microclimates engineered within structures to allow for terraformed expansion at a more leisurely pace.

The toxicity of other planets and the magnitude of the terraforming challenge, however, acts only to highlight how uniquely wondrous our own corner of the universe is. It is dangerous to believe an off-world home is a viable enough alternative to allow us rampantly to damage the paradise we're already so lucky to have.

Why visit Mars? As if inspiring technological progress or founding a potential off-Earth home isn't enough, Mars offers a chance to answer one of the most compelling reasons to travel into space in the first place – the opportunity to find life beyond Earth. The evidence for the presence of water, given our knowledge of evolution on Earth, makes it highly likely that life – whether fossilized or microbial – would be found on the Red Planet. It's expert research field work, the kind that even the most advanced rover would struggle to complete without a human at its side.

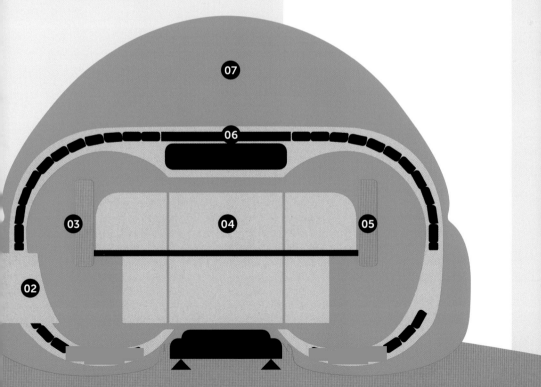

BIONIC IMP

WILL LET U

ENGINEER

THAT WOUL

MILLENNIA

LANTS

S

ABILITIES

D TAKE

TO EVOLVE

BIONIC IMPLANTS

From *The Six Million Dollar Man* to *RoboCop*, for decades pop culture has been obsessed with 'rebuilding' battered and broken heroes, installing a crushing robot arm here and super-speedy legs there.

As time has passed, the concept of a person having bionic (or artificial, mechanically enhanced) body parts has moved from fiction to scientific reality. We don't bat an eyelid when someone mentions that they have a pacemaker keeping their heart pumping healthily or a cochlear implant installed to aid their hearing.

These measures have historically been preventative or restorative treatments, working against a disease or condition. But, given the opportunity, would you replace a healthy limb or organ for a cybernetically upgraded alternative?

In the not-so-distant future, that may be a real possibility (an imperative, even), leap-frogging millennia of evolution with an off-the-shelf augmentation procedure acquired as easily as a facelift or tummy tuck.

Just looking at what's available today reveals the mind-boggling potential of tomorrow's improvements; paralympians running astounding race times with carbon-fibre-reinforced polymer prosthetics, and robotic hands that can 'see' what they're grasping even when the natural eye can't. Just as our bodies have evolved, so too will these technologies, but at a far more rapid rate than nature can manage.

We already have the means to restore some sight to the blind. The Argus II by California's Second Sight, for instance, combines a retinal implant with an external video camera mounted on a glasses frame to allow the visually impaired to see the edges of shapes in black and white. That's enough to cross a road unaided, or even read books with a large font size, sending a small, stimulating series of electrical pulses transmitted wirelessly to 60 electrodes implanted on the macula inside the eyeball – a direct line to the optic nerve.

Germany's Alpha IMS takes a different approach. It's a self-contained bionic eye connecting directly to the brain via 1,500 electrodes, with a built-in sensor that captures imagery from light passing into the eye, offering a sharper image than the Argus II (with the benefit of the owner being able to swivel the eyeball) at the expense of more invasive surgery.

This is just the start of what could be achieved. By tapping into the brain directly, there's the potential to expand the capabilities of human vision exponentially by teaching it to interpret signals from other external sensors. Our natural eyes can see only about 1% of the universe's light spectrum, but we already have gadgets that can see in heat-seeking infrared. Could these be features of the bionic eye of the future, complete with augmented-reality information overlays and camera-like zoom abilities?

Just as our bodies have evolved, so too will these technologies, but at a far more rapid rate than nature can manage.

BOOSTED BRAINS

Efforts are being made to enhance the brain itself in other ways too. Elon Musk's Neuralink has the ambitious (if yet unrealized) aim of joining our brains with artificial intelligence directly, letting us tap into networked information and bolster our memories to a photographic, capacious quality through brain-machine interfacing.

Neuralink discusses the brain in terms of two layers – the limbic system (responsible for emotion and memory) and the cortex (where analytic reasoning occurs). Musk's team of brain experts look to add a 'digital tertiary layer' to the mix that allows for high-bandwidth interfacing with AI, massively enhancing our intelligence levels and giving us the power to control external technologies through thought alone.

Where Neuralink's plan gets even more progressive is the fact that it aims to do all this without invasive surgery. Instead, the company looks to pioneer nanotech-enabled neural dust, made of thousands of tiny transmitters injected into the blood vessels that feed the brain, and a neural lace made of a fine lattice of sensors that could be embedded alongside it.

All this before considering the future-gazing, imaginative potential for fingers pre-programmed to make you a guitar-shredding rock god, lungs that could convert toxic gasses into breathable air, or a connected tongue that could simulate the dishes of history's greatest chefs from the comfort of your own home.

Bionic implants could let us engineer abilities that are way out of reach of humans, or would take incredibly specific conditions over millennia to evolve naturally. We could intercept and choose the direction of our evolution, letting technology dictate a new super-heroic path for our progress as a species. This is less a question of what bionics will let us do, but instead of what it will allow us to become, full stop.

Neuralink's team of brain experts look to add a 'digital tertiary layer' to the mix that allows for high-bandwidth interfacing with AI, massively enhancing our intelligence levels and giving us the power to control external technologies through thought alone.

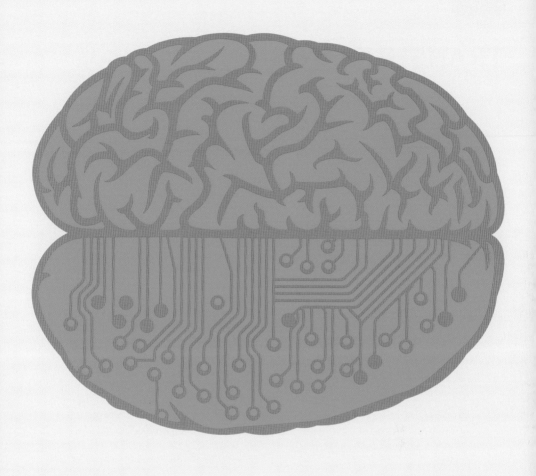

When bionics move away from being a restorative medical tool and become a lifestyle choice, how will that change our day-to-day lives? Will we look enviously on the echolocating ears of our augmented next-door neighbours in the same way we now lust over sports cars and designer clothes? Will we have to go under the knife to keep up with the Joneses? It may not simply be a case of fashion either. What will happen to those who can't afford body modifications, whose memories remain prone to natural human forgetfulness, and whose muscles tire at a rate evolution has dictated for millions of years? Will we also have engineered a new breed of elite person – those with the money to tap into experiences that would be impossible to understand for the 'un-augmented' human?

TRANSHUMANISM

Virtual worlds, artificial intelligences, nanorobots and bionic enhancements – we live in an age of unparalleled technological advancement. But why do we push forward with these developments, and what are we striving for?

If it is in order for humans to live better lives, then for many that aim would extend to living longer lives too. Improvements in living conditions and healthcare paired with technological refinements are continually lengthening the average human lifespan, which now sits at around 79 years. As we've come to learn, humans rarely rest on their laurels, instead continually building on the successes of their peers and predecessors. So what's better than a long life, well lived?

How about an eternal one?

It has been the promise of many religions over millennia, but as society increasingly puts its faith in science and technology, for some it has become the logical expectation for science and technology to ultimately 'solve' death.

Transhumanism is a movement that looks to technology and science to enhance human intellectual and physical capabilities, indefinitely postponing or circumventing the effects of ageing and finally eliminating our departure from life altogether. Trans-humanists don't see this unprecedented goal, driven not directly by nature but instead by human agency, as necessarily removed from the cycle of evolution, but simply our next stage in it. They believe human brilliance (and dominance) on Earth has let humanity become capable of defining its own progress, rather than being governed by the slow manifestation of the rules of Darwinism.

Transhumanists look either to the cumulative effects of multiple significant biological augmentations, and/or the creation of an entirely synthetic intelligence being built upon the foundations of our own brains to achieve this goal. But this endmost, superlative application of our technological progress will require not only the exploration of philosophical and moral questions, but also mastery of one of the human body's last remaining mysteries: how our minds work.

Whole brain emulation and mind uploading may sound like science fiction, but it's an area actively being worked on. The tissues of our bodies may degrade over time, but the thoughts that govern them are driven by electrical impulses and chemical changes – things that we could mimic and control, if only we could understand the underlying map of the brain through which they travel.

While our bodies are delicate and transient, there's potential to convert our thoughts and memories into data, with the hope that the personality and identity that accompanies them could live in some other substrate – whether organic or otherwise.

THE ETERNAL CONSCIOUSNESS

The average human brain contains 86 billion neurons, each with the ability to create thousands of connections to each other, with hundreds of trillions of synapses capable of exchanging information. This information travels and is processed at speeds that far exceed current computing methods, with – similar to the principles of quantum computing – neurons that can exist in states beyond simply 'on' or 'off'. While its ever-changing nature in no way makes it an analogous storage system, our understanding of how much information can be stored in the brain could be the equivalent of as much as 2.5 petabytes of digital storage space.

Barack Obama's administration saw the formation of the $100 million BRAIN initiative (Brain Research through Advancing Innovative Neurotechnologies), while the EU also committed €1.2 billion in a 10-year proposal to computationally simulate the brain too. Though these and similar programmes continue to make solid progress, aiding research into curing Alzheimer's and Parkinson's diseases, the vast intricacies of the brain still eludes us.

Which is not to say that has deterred the transhumanist faithful. The allure of an eternal life is so great that some have already committed their future deceased selves to the cause, paying huge sums of money so that their bodies (or in some cases, just their heads) can be cryogenically frozen, ready for a time in which they can be revived or have their minds restored through some future product of transhumanist research.

But is death really a problem? Would being able to exist forever change what it

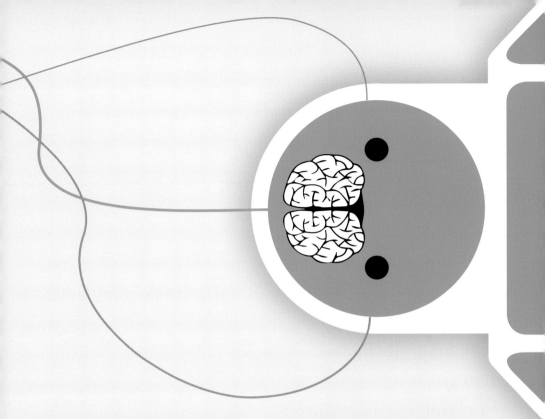

means to be human in the first place? We make decisions and weigh the importance of events in our lives today, knowing that they have a finite end date – it remains the only definite commonality between all human existence. Yet to embrace transhumanism would be to usher in the age of the posthuman universe. And 'universe' is a very important distinction to make over 'world'. With our consciousnesses disembodied, our identities branching, backed-up and potentially duplicated endlessly, we'd become able to explore the cosmos free from the limitations and fears associated with our fragile soft tissues and bones.

It would be easy to dismiss these ideas as far-fetched, or even mad. But history has a habit of creeping up on even the most liberal of minds. Ken Olsen, the founder of Digital

Equipment Corporation (which later merged with Compaq and Hewlett-Packard) famously had to eat his words after stating in 1977 that 'There is no reason why anyone would want a computer in their home.'

The technologies discussed in the pages of this book may hold the key not just to prolonging our lives, but also completely changing life as we know it. Could nanorobots scan our brains comprehensively and non-invasively extract what's stored within? Could brain-machine interfaces link us to networked quantum computers? Will whole-body bionics and advanced exoskeletons play host to our digitized brains? It may not happen within our finite lifetimes, but it's a future today's transhumanists hope to live long enough to see.

TOOLKIT

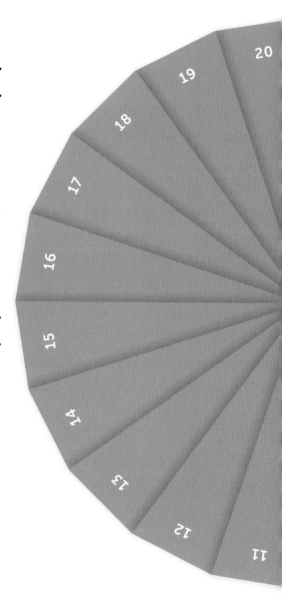

17

Quantum computers will use our knowledge
of quantum theory in order to ditch the bits
and bytes of traditional computers in favour
of quantum bits or 'qubits'. As these can exist
in many more states than the limited 'on' or
'off' values of a standard bit, they will allow
for machines massively more powerful than
the computers we have today. And while they
may remain the preserve of scientists and
researchers, a quantum computer's ability to
calculate problems quickly in parallel states
could be key to new cryptography, medical
and cosmological breakthroughs.

18

Terraforming refers to a number of
theoretical techniques that could
potentially be used to control and shape
the unwelcoming environments of
distant planets. A key part of any future
interplanetary colonization plans the
human race may have, it requires identifying
a planet within a star's suitably warm
'Goldilocks zone', analyzing the makeup
of its atmosphere and the strength of its
magnetosphere, and using anything from a
redirected asteroid to nuclear weapons in
order to recreate conditions to match what
we've evolved to live comfortably alongside
here on Earth. But there's no guarantee
these methods would work, highlighting just
how precious our own fragile planet is.

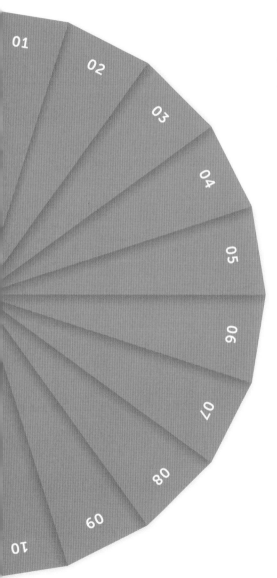

19

From pacemakers to hearing aids, we already live with some forms of bionic implants. But in the future, these won't necessarily just be prescribed as treatments, but as lifestyle choices to augment the capabilities of our bodies. From advanced vision to super-sensitive hearing to brain implants allowing you to keep a photographic memory of your life's every action, we'll be able to rebuild our bodies to suit our every need and desire.

20

With all the world's technologies – from advanced AIs to nanotechnologies and brain-machine interfaces – improving in ways we can sometimes scarcely predict, we may one day have the power to transcend our human forms through science and silicon. The movement known as transhumanism looks forward to a time when we'll be able to augment our bodies (or replace them altogether) in such a way that our minds will live on eternally in a digital state, ushering in a new posthuman age.

FURTHER LEARNING

READ

Homo Deus: a Brief History of Tomorrow
Yuval Noah Harari (Harvill Secker, 2016)

To be a Machine:
Adventures Among Cyborgs, Utopians,
Hackers, and the Futurists Solving the
Modest Problem of Death
Mark O'Connell (Granta Books, 2017)

How to Create a Mind:
The Secret of Human Thought Revealed
Ray Kurzweil
(Gerald Duckworth & Co Ltd, 2014)

Superintelligence:
Paths, Dangers, Strategies
Nick Bostrom (OUP Oxford, 2014)

Computing with Quantum Cats:
From Colossus to Qubits
John Gribbin (Black Swan, 2015)

WATCH

Fixed: the Science/Fiction of Human
Enhancement
This documentary follows the lives of
five people living with disabilities while
also working within areas relating to
transhumanism. It's a great exploration of
how, for better and worse, transhumanist
technologies and body enhancements can
affect lives and a person's sense of identity.

STUDY

Geology with Planetary Science
The University of Manchester, UK

Biomedical Science BSc
King's College London, UK

VISIT

**Alcor Life Extension Foundation, Scottsdale,
Arizona, USA**
This is the world's leading public-facing
cryogenics facility. It freezes the bodies of
members with the aim of caring for them
until a future arrives in which they can be
revived, and their lives prolonged. Free tours
can be booked every week.

EPILOGUE

This book began by stating that your world is changing. It ends with the fact that you too will change alongside it.

Under the right circumstances – political, social and economic – technological advances can bring about change to the world very quickly. But the next technological revolution may urge us to anticipate change not just in the world around us, but in the very definition of our existence.

As we look to solve what we see as limitations, technology's future seems set on amplifying and augmenting our bodies and minds. Autonomous transport and AI-controlled homes and workforces may make us more efficient than ever, but leave our physical bodies in need of catching up with this new world we have created.

Today we choose the model of our smartphones; in the distant tomorrow, could we choose a model for our selves? Should the act of removing our minds from our physical bodies ever be realized, will our flesh and bones go the way of the floppy disc and the VHS? Would our current physical forms become outmoded (albeit biological) hardware, to be superseded by something not yet realized – the traces of iron, copper and silicon already found in our bodies amplified?

Defining which version of your digitally disembodied 'self' was the authentic 'you', potentially modified, copied and restarted infinitely, would be a philosophical abyss to navigate, but a society that could develop such technological scenarios would most likely have previously redefined the nature and purpose of the self anyway.

Imagine existing anywhere you liked, or in multiple places at once, time malleable and irrelevant, so long as there was some suitable substrate in which you could house your uploaded consciousness. You could share thoughts in the same way that today we share data, transmitted to the whole posthuman race in an instant. It's hard to fathom, but removing our inner selves from our physical forms could, literally, unite mankind in ways currently unimaginable. It

Today we choose the model of our smartphones — in the distant tomorrow, could we choose a model for our selves?

would be unlike any life history has ever known.

It may seem far-fetched today, but like all the technological advancements discussed, it is the impossible dreams that make small steps happen. Just as the smartphone has put the world in the palm of our hands, we may see the blockchain revolutionize security systems, nuclear fusion lead to a clean energy breakthrough, and nanotechnology and biological augmentations allow for ever-more ambitious forays among the stars. Future innovations will continue to surprise and surpass our expectations.

There is no fate but what we make for ourselves. We live in an age of unprecedented change and technological upheaval. But that technology also puts all of civilization's combined knowledge at our fingertips, connecting us with the greatest minds across the globe, with devices that empower us to think bigger, bolder and brighter than ever before.

Your world is changing — and it's yours to change.

BUILD +
BECOME

BIBLIOGRAPHY

Practical Augmented Reality: a Guide to the Technologies, Applications and Human Factors for AR and VR
Steve Aukstakalnis
(Addison-Wesley Professional, 2016)

A Piece of the Sun: the Quest for Fusion Energy
Daniel Clery
(Gerald Duckworth & Co. Ltd, 2013)

The Master Algorithm: How the Quest for the Ultimate Learning Machine Will Remake Our World
Pedro Domingos (Penguin, 2015)

Driverless: Intelligent Cars and the Road Ahead
Hod Lipson and Melba Kurman
(MIT Press, 2016)

The Quantified Self
Deborah Lupton
(Polity Press, 2016)

The New Net Zero: Leading-Edge Design and Construction of Homes and Buildings for a Renewable Energy Future
William Maclay
(Chelsea Green Publishing, 2014)

Trackers: How Technology is Helping Us Monitor and Improve Our Health
Richard MacManus
(Rowman and Littlefield Publishers, 2015)

Elon Musk: How the Billionaire CEO of SpaceX and Tesla Is Shaping Our Future
Ashlee Vance
(Virgin Books, 2016)

How We'll Live on Mars
Stephen Petranek
(Simon & Schuster, 2015)

The Case for Mars: the Plan to Settle the Red Planet and Why We Must
Robert Zubrin
(Free Press, 2011)

Wearable Robots: Biomechatronic Exoskeletons
José L. Pons
(Wiley, 2008)

Engines of Creation: the Coming Era of Nanotechnology
K Eric Drexler
(Anchor, 1986)

Near-Earth Objects: Finding Them Before They Find Us
David K. Yeomans
(Princeton University Press, 2016)

Future Crimes
Marc Goodman
(Doubleday Books, 2015)

We Are Anonymous
Parmy Olson
(William Heinemann, 2013)

Digital Gold: the Untold Story of Bitcoin
Nathaniel Popper
(Penguin, 2016)

Kill Chain: Rise of the High-Tech Assassins
Andrew Cockburn
(Picador USA, 2016)

*Dark Territory: the Secret History of
Cyber War*
Fred Kaplan
(Simon & Schuster, 2016)

Homo Deus: a Brief History of Tomorrow
Yuval Noah Harari
(Harvill Secker, 2016)

Human Enhancement
Julian Savulescu (Editor), Nick Bostrom
(Series Editor)
(Oxford University Press, 2011)

Sapiens: a Brief History of Humankind
Yuval Noah Harari (Vintage, 2015)

*To Be a Machine: Adventures Among
Cyborgs, Utopians, Hackers, and the
Futurists Solving the Modest Problem
of Death*
Mark O'Connell
(Granta Books, 2017)

How to Create a Mind
Ray Kurzweil
(Gerald Duckworth & Co Ltd, 2014)

Transcend: 9 Steps to Living Well Forever
Ray Kurzweil, Terry Grossman
(Rodale Press, 2010)

*Superintelligence: Paths, Dangers,
Strategies*
Nick Bostrom
(OUP Oxford, 2014)

*Computing with Quantum Cats:
From Colossus to Qubits*
John Gribbin
(Black Swan, 2015)

PODCASTS

FutureProofing BBC Radio 4,
Leo Johnson and Timandra Harkness

Singularity.FM., Nikola Danaylov

At BUILD+BECOME we believe in building knowledge that helps you navigate your world.

Our books help you make sense of the changing world around you by taking you from concept to real-life application through 20 accessible lessons designed to make you think. Create your library of knowledge.

BUILD +
BECOME

www.buildbecome.com
buildbecome@quarto.com

@buildbecome
@QuartoExplores

Through a series of 20 practical and effective exercises, all using a unique visual approach, Michael Atavar challenges you to open your mind, shift your perspective and ignite your creativity. Whatever your passion, craft or aims, this book will expertly guide you from bright idea, through the tricky stages of development, to making your concepts a reality.

We often treat creativity as if it was something separate from us – in fact it is, as this book demonstrates, incredibly simple: creativity is nothing other than the very core of 'you'.

Michael Atavar is an artist and author. He has written four books on creativity – *How to Be an Artist, 12 Rules of Creativity, Everyone Is Creative* and *How to Have Creative Ideas in 24 Steps – Better Magic*. He also designed (with Miles Hanson) a set of creative cards *'210CARDS'*.

He works 1-2-1, runs workshops and gives talks about the impact of creativity on individuals and organisations.
www.creativepractice.com

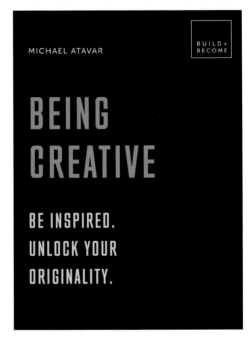

MICHAEL ATAVAR

BUILD+
BECOME

BEING CREATIVE

BE INSPIRED. UNLOCK YOUR ORIGINALITY.

CREATIVITY BEGINS WITH YOU.

Using a unique, visual approach to explore the science of behaviour, *Read People* shows how understanding why people act in certain ways will make you more adept at communicating, more persuasive and a better judge of the motivations of others.

The increasing speed of communication in the modern world makes it more important than ever to understand the subtle behaviours behind everyday interactions. In 20 dip-in lessons, Rita Carter translates the signs that reveal a person's true feelings and intentions and exposes how these signals drive relationships, crowds and even society's behaviour. Learn the influencing tools used by leaders and recognise the fundamental patterns of behaviour that shape how we act and how we communicate.

Rita Carter is an award-winning medical and science writer, lecturer and broadcaster who specialises in the human brain: what it does, how it does it, and why. She is the author of *Mind Mapping* and has hosted a series of science lectures for public audience. Rita lives in the UK.

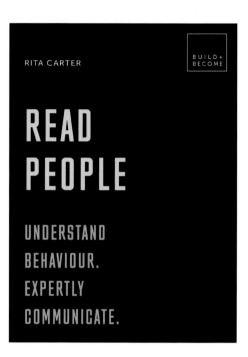

RITA CARTER

BUILD +
BECOME

READ PEOPLE

UNDERSTAND
BEHAVIOUR.
EXPERTLY
COMMUNICATE.

CAN YOU SPOT A LIE?

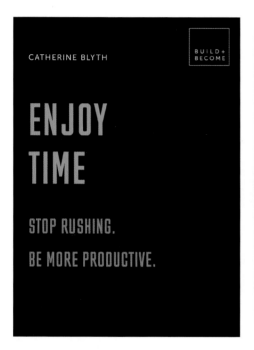

CATHERINE BLYTH

BUILD+
BECOME

ENJOY TIME

STOP RUSHING.

BE MORE PRODUCTIVE.

October 2018

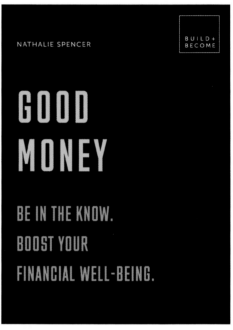

NATHALIE SPENCER

BUILD+
BECOME

GOOD MONEY

BE IN THE KNOW.

BOOST YOUR
FINANCIAL WELL-BEING.

October 2018

Using a unique, visual approach to explore philosophical concepts, Adam Ferner shows how philosophy is one of our best tools for responding to the challenges of the modern world.

From philosophical 'people skills' to ethical and moral questions about our lifestyle choices, philosophy teaches us to ask the right questions, even if it doesn't necessarily hold all the answers. With 20 dip-in lessons from history's great philosophers alongside today's most pioneering thinkers, this book will guide you to think deeply and differently.

Adam Ferner has worked in academic philosophy both in France and the UK – but it's philosophy *outside* the academy that he enjoys the most. In addition to his scholarly research, he writes regularly for *The Philosophers' Magazine*, works at the Royal Institute of Philosophy and teaches in schools and youth centres in London.

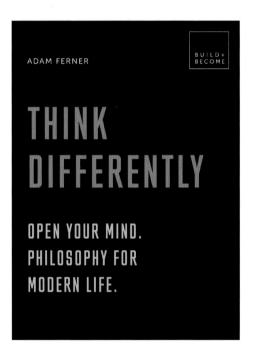

ADAM FERNER

BUILD + BECOME

THINK DIFFERENTLY

OPEN YOUR MIND.
PHILOSOPHY FOR
MODERN LIFE.

PHILOSOPHY IS ABOUT OUR LIVES AND HOW WE LIVE THEM.

04
BUILD +
BECOME

THE LIMITS OF LOYALTY

When was the last time you felt betrayed? Have a good think ... got it? It's a horrible feeling, isn't it? Really gut-churning. It's tremendously upsetting to discover someone you trusted has abused that trust. The connection you thought you had has been severed, perhaps irrevocably, and the time, effort and love that's gone into a relationship has all gone to waste.

Betrayal comes in many forms. You can betray your partner by hooking up with others; you can betray your country by selling its secrets; you can betray your friends by talking dirt about them behind their backs. You can betray yourself too, by failing to live up to the principles you tell yourself to abide by.

The harmful effects of these betrayals can be far-reaching. If you betray your partner, you imperil your relationship. You also imperil future relationships – if you've cheated on someone you love, who's to say you won't do so again? You damage your status as a 'trustworthy' individual. Betrayal can be seen as an indication of weakness and inconstancy.

It's because of these harms that we have so many damning terms for betrayers: fair-weather friends, turncoats, traitors, snitches, sneaks, weasels, rats ... And it's because of these harms that we often figure loyalty is a good thing.

The 19th-century American philosopher Josiah Royce wrote in his book, *The Philosophy of Loyalty* (1908), that loyalty is 'the willing and practical and thorough-going devotion of a person to a cause'. We can be loyal to ideals and institutions. More recently, the moral philosopher, Marcia Brown, has pointed out that we more typically take 'loyalty' to refer to a relationship between *persons*. You can be loyal to your partner if, despite being attracted to someone else, you nevertheless remain faithful. You can be loyal to your friend if, despite pressure to sell them out, you keep schtum. We tend to prize these kinds of actions and typically label loyalty as 'a virtue'.

Loyalty offers security. If you're the nervous type, your partner's loyalty will offer considerable comfort. And loyalty is good for both you and the recipient. If you're loyal to something (a football club, for example), your relationship to that thing is enhanced. You identify with the institution or person or principle that you're loyal to. If I'm loyal to my family, and put the needs of my family before my own, it's because, in a way, I see my own needs as inextricably tied up with theirs.

Loyalty, then, seems to be a good thing. But seeming is different from being. Is loyalty so certain a virtue?

32

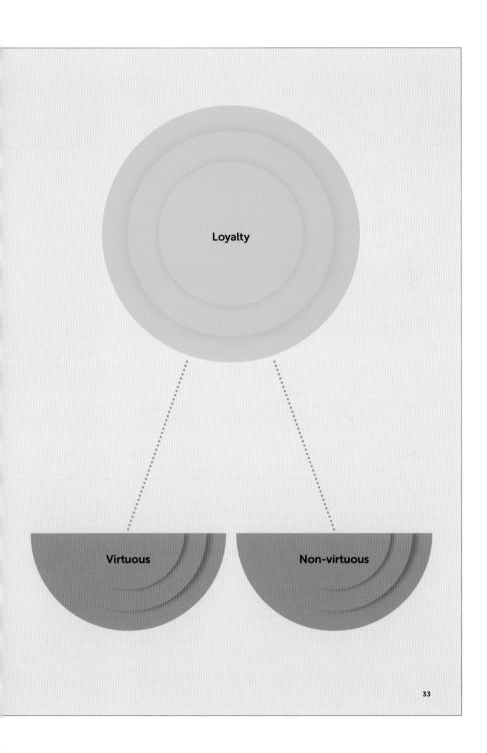

Loyalty

Virtuous

Non-virtuous

33

04
BECOME

LOYALTY

DISLOYALTY

THE VALUES OF DISLOYALTY

'Loyalty confines you to accepted opinions; loyalty forbids you to comprehend sympathetically your dissident fellows...' *Graham Greene*

'The first thing I want to teach is disloyalty till they get used to disusing that word loyalty as representing a virtue. This will beget independence...' *Mark Twain*

There are, as these quotations attest, a few folk who are slightly more circumspect about the so-called virtue of loyalty. The general thought, articulated by the novelists Twain and Greene, is that loyalty is restrictive. It undermines your independence. It's a form of control. The examples discussed above show us its positive effects, but there are plenty of cases where it can have slightly more dubious results.

Consider Akosua. She has been working hard at her job for a couple of years and has swiftly risen to managerial level. Perhaps she has been helped in some minimal way by one of the partners in the firm: a kindly old soul called John. One day Akosua discovers that John has been paying for fancy dinners with the company credit card. It's wrong, and she knows it. So should she report him? The Johns of this world typically rely on a 'sense of loyalty' to keep people less powerful than themselves quiet. Would disloyalty in this situation really be a bad thing?

It's easy to imagine even more troubling cases where people in power use loyalty to silence their employees and co-workers.

34

As Marcia Baron points out in *The Moral Status of Loyalty* (1984), there are numerous instances of disloyalty having beneficial effects and undermining objectionable institutions. Remember, it's the concept of loyalty – to one's queen, or lord – which supported the hugely unequal feudal society. And it's disloyalty, or 'whistle blowing', that means that otherwise well-protected industries that benefit from, for example, slave labour, can be brought to justice. Disloyalty to presidents, queens and company bosses can bring to light hidden abuses of power.

Loyalty, says Baron, can stand as an obstacle to justice. It's a dissuasive force that encourages you to leave certain things unquestioned. This thoroughgoing commitment to a person or cause overrides your ability to critique the objects of your loyalty effectively. The consequences of such loyalty can be horrendous. Take, for example, the German soldiers who were unthinkingly loyal to the Third Reich, or the British citizens who gave unswerving loyalty to the empire and its exploitative colonial activities.

Sure, loyalty can prevent you from doing bad things. It can stop you from cheating on your partner. But loyalty also protects others when they do bad things. It allows businesses to continue in their dodgy dealings, and leaves individuals open to exploitation. The effects of loyalty can be both positive and negative – and, interestingly, the same is true for disloyalty. So if loyalty's a virtue, maybe disloyalty can be one too?

ACKNOWLEDGEMENTS

Special thanks go to Lucy Warburton, for giving me the opportunity to peer into the future on the page, Stuart Tolley, for his fantastic illustration and design work, Victoria Marshallsay, for wrangling my words; Duncan Geere, for getting my facts straight, Andy Roberts, for a crash course in nuclear fusion; and Patrick Goss and the entire TechRadar.com team, for their support and generosity while sharing their unmatched expertise. Extra-special thanks go to Kevin Lynch, for getting me into all this mad stuff in the first place. The beers are on me.

Gerald Lynch is a technology and science journalist, and is currently Senior Editor of technology website TechRadar. Previously Editor of websites Gizmodo UK and Tech Digest, he has also written for publications such as *Kotaku* and *Lifehacker*, and is a regular technology pundit for the BBC. Gerald was on the judging panel for the James Dyson Award. He lives in London.